I Can Do Math Practice Problems GRADE 2

Volume 2: Units 5 - 9

This book belongs to:

_____

I Can Do Math Practice Problems GRADE 2

Volume 2: Units 5 - 9

Dr. Christine Scafidi

Copyright © 2025 by CKingEducation, Inc.

All rights reserved.

This publication is protected under United States and international copyright law. No part of this book may be reproduced, stored in a retrieval system, shared digitally, distributed, or transmitted in any form or by any means, electronic, mechanical, photocopying, recording, scanning, or otherwise, without the express prior written permission of the publisher. This includes, but is not limited to, uploading to shared drives, copying for multiple students, or use in multiple classrooms.

This book is licensed for use by a single teacher in a single classroom only. Reproduction or redistribution beyond the original purchase violates copyright law and publisher licensing terms.

Brief excerpts may be quoted in reviews or scholarly works in accordance with Section 107 of the U.S. Copyright Act (fair use), provided proper attribution is given.

While every effort has been made to ensure the accuracy of the content at the time of publication, the authors and publisher assume no responsibility for errors, omissions, or for any outcomes resulting from the use of this material.

Published by CKingEducation

CKingEducation

To contact CKingEducation or the authors about speaking, workshops, or ordering books in bulk, visit www.ckingeducation.com.

ISBN: 978-1-968264-01-7

Lead writer: Dr. Christine Scafidi
Editor: John Sasko
Series editor and book design: Christine King

Printed in the United States of America

Table of Contents

Track what you have done by checking off items.

I did it!	Unit/Practice/I Can Statement	Page
	Unit 5 Practice 1: I can model numbers to 100 using base-ten representations.	16
	Unit 5 Practice 2: I can model hundreds using base-ten blocks.	18
	Unit 5 Practice 3: I can write three-digit numbers using base-ten models.	20
	Unit 5 Practice 4: I can make base-ten representations of three-digit numbers.	22
	Unit 5 Practice 5: I can write three-digit numbers using expanded form.	24
	Unit 5 Practice 6: I can write two-digit and three-digit numbers in words.	26
	Unit 5 Practice 7: I can represent three-digit numbers in different ways.	28
	Unit 5 Practice 8: I can count tens and hundreds on a number line.	30
	Unit 5 Practice 9: I can compare three-digit numbers.	32
	Unit 5 Practice 10: I can use base-ten representations to compare numbers.	34
	Unit 5 Practice 11: I can use place value to compare three-digit numbers.	36
	Unit 5 Practice 12: I can order numbers from least to greatest.	38
	Unit 5 Practice 13: I can order numbers from greatest to least.	40
	Unit 5 Practice 14: I can count collections to 1,000.	42
	Unit 5 Fluency Practice	44

Table of Contents

Track what you have done by checking off items.

I did it!	Unit/Practice/I Can Statement	Page
	Unit 6 Practice 1: I can name and list attributes of flat shapes.	50
	Unit 6 Practice 2: I can draw flat shapes.	52
	Unit 6 Practice 3: I can use a 1 inch dot grid to draw shapes.	54
	Unit 6 Practice 4: I can name and describe solid shapes.	56
	Unit 6 Practice 5: I can describe and compare flat shapes and solid shapes.	58
	Unit 6 Practice 6: I can make a flat shape using smaller shapes.	60
	Unit 6 Practice 7: I can partition shapes into equal pieces.	62
	Unit 6 Practice 8: I can partition shapes into halves, thirds and fourths.	64
	Unit 6 Practice 9: I can show equal shares using different shapes.	66
	Unit 6 Practice 10: I can work with equal shares.	68
	Unit 6 Practice 11: I can use halves and quarters to tell time on an analog clock.	70
	Unit 6 Practice 12: I can tell time using an analog clock.	72
	Unit 6 Practice 13: I can make a schedule to show a.m. or p.m.	74
	Unit 6 Practice 14: I can show time on an analog clock and on a digital clock.	76
	Unit 6 Practice 15: I can identify coin names and coin values.	78

I Can Do Math Practice Problems, Grade 2   © 2025 www.ckingeducation.com

Table of Contents

Track what you have done by checking off items.

I did it!	Unit/Practice/I Can Statement	Page
	Unit 6 Practice 16: I can compare the value of coin collections.	80
	Unit 6 Practice 17: I can find coin combinations that have the value of one dollar.	82
	Unit 6 Practice 18: I can solve word problems using money.	84
	Unit 6 Practice 19: I can solve story problems using units of a dollar.	86
	Unit 6 Practice 20: I can solve and write story problems using money.	88
	Unit 6 Practice 21: I can make designs with pattern blocks.	90
	Unit 6 Fluency Practice	92
	Unit 7 Practice 1: I can use number lines to add, subtract and compare numbers.	98
	Unit 7 Practice 2: I can use base-ten blocks to model three-digit numbers.	100
	Unit 7 Practice 3: I can use models to add and subtract.	102
	Unit 7 Practice 4: I can use expanded form to add or subtract.	104
	Unit 7 Practice 5: I can use place value to find the next hundred	106
	Unit 7 Practice 6: I can compose a ten when adding.	108
	Unit 7 Practice 7: I can compose a ten or compose a hundred when adding.	110
	Unit 7 Practice 8: I can compose a ten and a hundred to add 3-digit numbers.	112

Table of Contents

Track what you have done by checking off items.

I did it!	Unit/Practice/I Can Statement	Page
	Unit 7 Practice 9: I can add two 3-digit numbers.	114
	Unit 7 Practice 10: I can find sums to 1,000 using different strategies.	116
	Unit 7 Practice 11: I can find the sum of two 3-digit numbers.	118
	Unit 7 Practice 12: I can decompose a ten to subtract.	120
	Unit 7 Practice 13: I can decompose a hundred to subtract.	122
	Unit 7 Practice 14: I can decompose a ten or a hundred to subtract.	124
	Unit 7 Practice 15: I can decompose a ten and a hundred to subtract.	126
	Unit 7 Practice 16: I can use base-ten numerals to subtract 3-digit numbers.	128
	Unit 7 Practice 17: I can use a number line to subtract.	130
	Unit 7 Practice 18: I can use different strategies to subtract.	132
	Unit 7 Fluency Practice	134
	Unit 8 Practice 1: I can use objects to make equal groups.	140
	Unit 8 Practice 2: I can make pairs.	142
	Unit 8 Practice 3: I can identify even and odd groups of objects.	144
	Unit 8 Practice 4: I can decompose a number to show even or odd.	146

I Can Do Math Practice Problems, Grade 2 © 2025 www.ckingeducation.com

Table of Contents

Track what you have done by checking off items.

I did it!	Unit/Practice/I Can Statement	Page
	Unit 8 Practice 5: I can find patterns using even numbers and odd numbers.	148
	Unit 8 Practice 6: I can show sums with more than two addends.	150
	Unit 8 Practice 7: I can count rows in an array.	152
	Unit 8 Practice 8: I can count columns in an array.	154
	Unit 8 Practice 9: I can use an expression to model an array.	156
	Unit 8 Practice 10: I can draw arrays and write expressions.	158
	Unit 8 Practice 11: I can write an equation to model an array.	160
	Unit 8 Practice 12: I can fill rectangles with equal size squares to model arrays.	162
	Unit 8 Practice 13: I can find sums and differences and partition rectangles.	164
	Unit 8 Fluency Practice	166
	Unit 9 Practice 1: I can add and subtract within 20.	172
	Unit 9 Practice 2: I can use addition or subtraction to find sums and differences.	174
	Unit 9 Practice 3: I can add or subtract on a grid to find distance.	176
	Unit 9 Practice 4: I can use measurement data to make a line plot.	178
	Unit 9 Practice 5: I can show 3-digit numbers in different ways.	180

Table of Contents

Track what you have done by checking off items.

I did it!	Unit/Practice/I Can Statement	Page
	Unit 9 Practice 6: I can write equations using equal expressions.	182
	Unit 9 Practice 7: I can use different strategies to add and subtract within 1,000.	184
	Unit 9 Practice 8: I can add and subtract within 100 using mental math.	186
	Unit 9 Practice 9: I can solve different types of story problems.	188
	Unit 9 Practice 10: I can write a question from a story problem.	190
	Unit 9 Practice 11: I can connect tape diagrams to story problems and equations.	192
	Unit 9 Practice 12: I can solve and write one-step story problems.	194
	Unit 9 Practice 13: I can solve and write two-step story problems.	196
	Unit 9 Fluency Practice	196

I Can Do Math Practice Problems, Grade 2 © 2025 www.ckingeducation.com

Welcome to 2nd Grade!

Doing these practice problems will support you to better understand the skills and concepts you are learning.

Unit 5
Numbers to 1,000

In this unit, we will learn how to read, write, and work with numbers up to 1,000. We will use place value and number lines to help us understand, compare, and order these numbers.

Model the Three-Digit Number in Many Ways

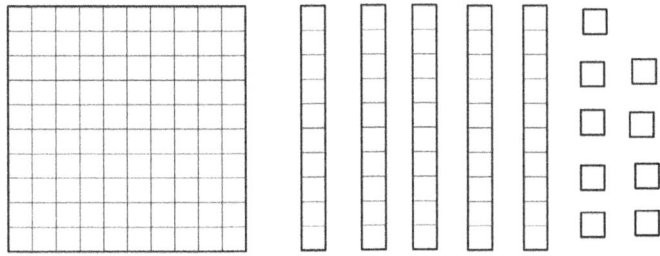

Three-digit number: 159
In Words: one hundred fifty-nine
In Expanded Form: 100 + 50 + 9
Using Base-Ten Numerals: 1 hundred, 5 tens, 9 ones

Compare Numbers Using a Number Line

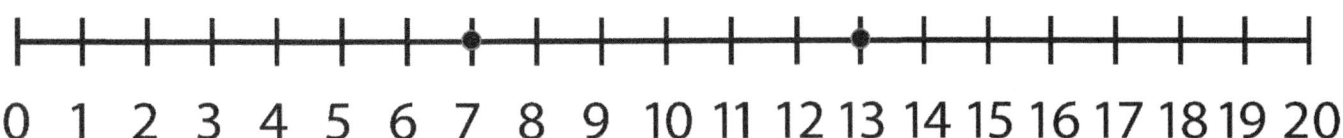

13 > 7

Unit Vocabulary

Numbers to 1,000

Use this space to visualize the math vocabulary for this unit.

Word or Phrase	Example or Attributes	Visual Reminder

Unit Models & Strategies

Numbers to 1,000

Use this space to visualize the math models and strategies for this unit.

Model or Strategy	This is a...	It is used to...

Affirmation: Acting out math problems helps me understand what I am doing.

Name: ______________________ Date: ____________

Unit 5 Practice 1: I can model numbers to 100 using base-ten representations.

1) How many? Write the 2-digit number.

______

1.NBT.2

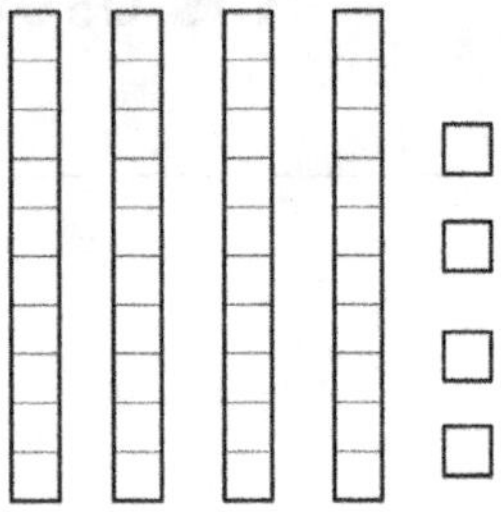

2) How many? Write the 2-digit number.

______

1.NBT.2

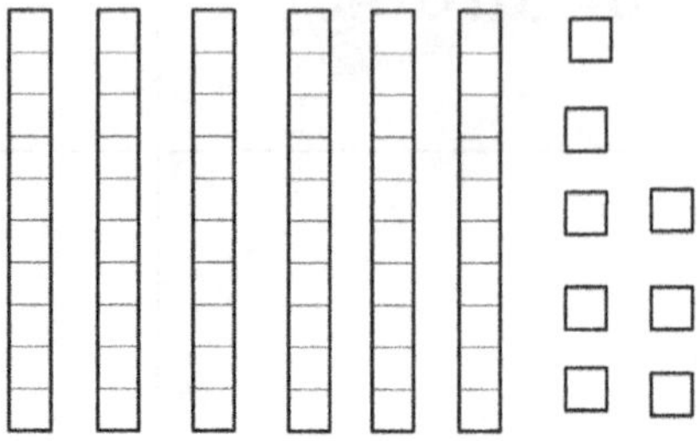

3) Represent 97 using base-ten blocks.

1.NBT.2c

4) How many tens?

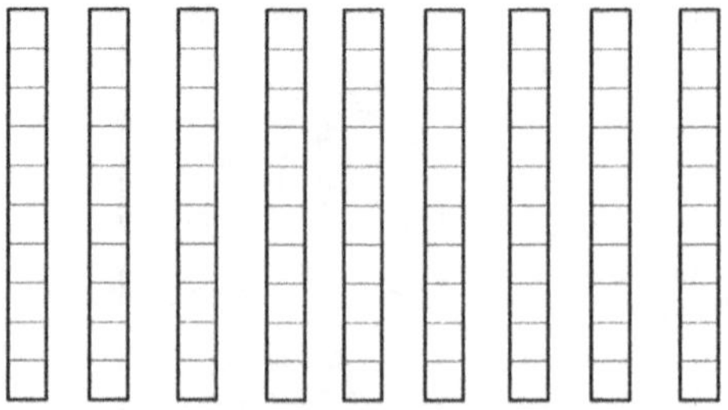

There are ______ tens.

Write the two-digit number ______.

1.NBT.2c

Do you understand? **?** ✓

Affirmation: Acting out math problems helps me understand what I am doing.

Name: _____ Date: _____

Unit 5 Practice 1: I can model numbers to 100 using base-ten representations.

5) How many more to make 100?

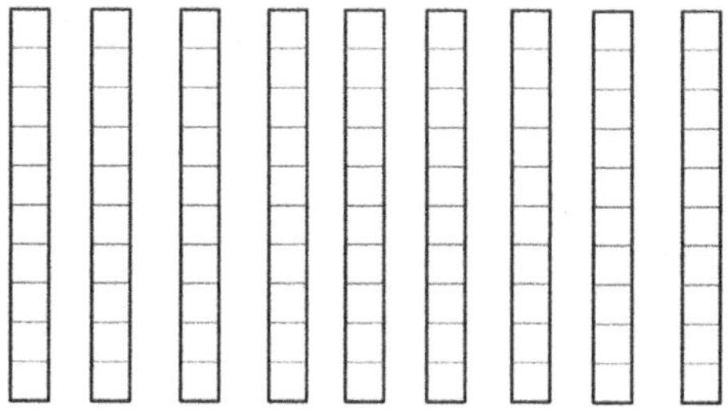

There are _____ tens.

I need _____ more ten to make 100.

2.NBT.1

6) Draw the base-ten blocks needed to make 100.

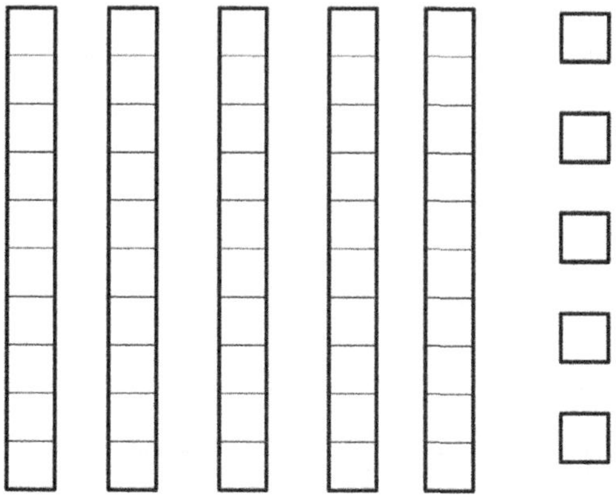

2.NBT.1

Do you understand?   ?   ✓

Affirmation: Sketching out my thinking helps me see a problem more clearly.

Name: _____ Date: _____

Unit 5 Practice 2: I can model hundreds using base-ten blocks.

1) Count on by tens.

   10, _____, _____, _____, _____, _____, _____, _____, _____, 100

   130, _____, _____, _____, _____, _____, _____, _____, 210

   240, _____, _____, _____, _____, _____, _____, _____, 320

   2.NBT.2

2) Count by tens to 100 using base-ten blocks.

   10 _____ _____ _____ _____ _____ _____ _____ _____ _____ 100

   2.NBT.2

3) How many? Circle the three-digit number the base-ten blocks represent.

   300   120   210

   2.NBT.1

Do you understand?   ?   ✓

Affirmation: Sketching out my thinking helps me see a problem more clearly.

Name: _____ Date: _____

Unit 5 Practice 2: I can model hundreds using base-ten blocks.

4) Count by hundreds.

   100, 200, _____, _____, _____, _____, 700

   300, _____, _____, _____, _____, _____, 900

   400, _____, _____, _____, _____, _____, 1,000

   2.NBT.2

5) Count by hundreds using base-ten blocks.

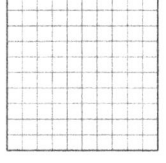 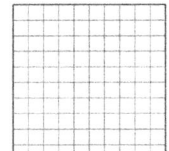 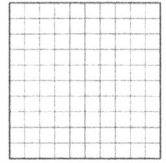 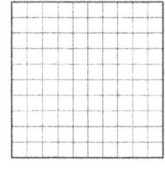 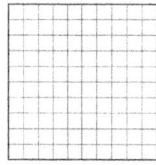 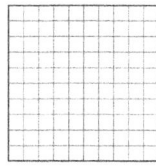

   100   _____   _____   _____   _____   _____

   2.NBT.2

6) How many? Circle the three-digit number the base-ten blocks represent.

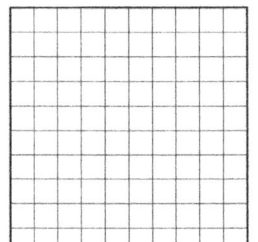 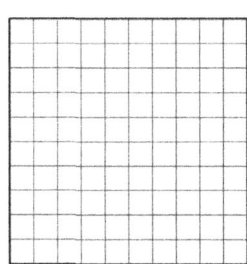 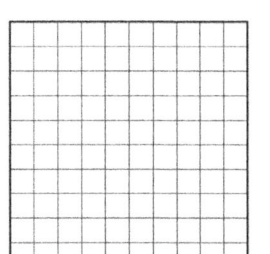 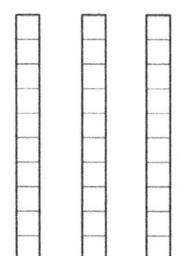

   303

   333

   330

   2.NBT.2

Do you understand?   ?   ✓

Affirmation: Making mistakes is how I learn new things.

Name: _____ Date: _____

Unit 5 Practice 3: I can write three-digit numbers using base-ten models.

1) How many tens? How many ones?

There are _____ tens.

There are _____ ones.

The two-digit number is _____.

1.NBT.2

2) Show with base-ten blocks.

6 tens   9 ones

The two-digit number is _____.

1.NBT.2

3) How many hundreds? How many tens? How many ones?

There are _____ hundreds.

There are _____ tens.

There are _____ ones.

The three-digit number is _____.

2.NBT.3

Do you understand?   ?   ✓

Affirmation: Making mistakes is how I learn new things.

Name: _____ Date: _____

Unit 5 Practice 3: I can write three-digit numbers using base-ten models.

4) How many hundreds? How many tens? How many ones?

There are _____ hundreds.

There are _____ tens.

There are _____ ones.

The three-digit number is:

2.NBT.3

5) How many hundreds? How many tens? How many ones?

There are _____ hundreds.

There are _____ tens.

There are _____ ones.

The three-digit number is:

2.NBT.3

Do you understand?   ?   ✓

I Can Do Math Practice Problems, Grade 2

Affirmation: I can help others by asking questions.

Name: _____ Date: _____

Unit 5 Practice 4: I can make base-ten representations of three-digit numbers.

1) Match the number that makes each equation true.

20 + 5 = ☐   430

4 + 30 + 100 = ☐   134

60 + 1 + 200 = ☐   25

30 + 400 = ☐   261

2.NBT.1

2) Circle true 👍 or false 👎 for each equation.

20 + 100 + 6 = 126   3 + 1 + 5 = 315

483 = 3 + 800 + 400   500 + 40 + 9 = 549

2.NBT.1

Do you understand?   ?   ✓

Affirmation: I can help others by asking questions.

Name: _____ Date: _____

Unit 5 Practice 4: I can make base-ten representations of three-digit numbers.

3) Draw the three digit number using base-ten representations.

5 hundreds, 6 ones, 3 tens   The three-digit number is _____

2.NBT.3

4) Draw the three digit number using base-ten representations.

4 ones, 0 tens, 6 hundreds   The three-digit number is _____

2.NBT.3

Do you understand?   ?   ✓

Affirmation: I can do harder math problems by acting out the story.

Name: _____   Date: _____

Unit 5 Practice 5: I can write three-digit numbers using expanded form.

1) **Sample:** Three-digit numbers modeled in expanded form.
   Expanded form is the sum of the value of each digit.

   Three digit number: 374

   The digit 3 has the value of 300.
   The digit 7 has the value of 70.
   The digit 4 has the value of 4.

   Expanded form: 374 = 300 + 70 + 4

   2.NBT.3

2) Find the the value of each digit in the number 582.
   Write the three-digit number 582 in **expanded form**.

   The digit 5 has the value of _____.

   The digit 8 has the value of _____.

   The digit 2 has the value of _____.

   Expanded form: _____

   2.NBT.3

3) Write the missing numbers to show expanded form.

   543 = 500 + _____ + 3   851 = 1 + _____ + 50

   200 + 60 + 7 = _____   70 + 4 + 100 = _____

   439 = _____ + 9 + _____   382 = _____ + _____ + _____

   2.NBT.3

Do you understand?   ?   ✓

Affirmation: I can do harder math problems by acting out the story.

Name: _____ Date: _____

Unit 5 Practice 5: I can write three-digit numbers using expanded form.

4) Use the base-ten blocks to write the three-digit number.
   Show the number in **expanded form**.

   Three digit number: _____

   Expanded Form: _____

   2.MD.3

5) Use the base-ten blocks to write the three-digit number.
   Show the number in **expanded form**.

   Three digit number: _____

   Expanded Form: _____

   2.MD.3

Do you understand?   ?   ✓

Affirmation: I am a problem solver.

Name: _____   Date: _____

Unit 5 Practice 6: I can write two-digit and three-digit numbers in words.

1) Writing numbers from number words.

    five _____   twelve _____   four _____   thirteen _____

    two _____   eleven _____   nine _____   six _____

    eight _____   seven _____   ten _____   zero _____

    three _____   fifteen _____   one _____   fourteen _____

2.NBT.3

2) Count by tens in words.

    ten, _____, _____, _____,

    fifty, _____, _____, _____,

    _____, one hundred.

2.NBT.3

3) Write the two-digit or three-digit number:

    sixty-seven _____   four hundred twelve _____

    two hundred five _____   twenty-nine _____

    nine hundred thirty _____   ninety-eight _____

2.NBT.3

Do you understand?   ?

Affirmation: I am a problem solver.

Name: _____   Date: _____

Unit 5 Practice 6: I can write two-digit and three-digit numbers in words.

4) Write each three-digit number in words.

　　Three-digit number: 362

　　In words: _____ hundred sixty-_____

　　Three-digit number: 581

　　In words: _____ hundred _____ -one

　　Three-digit number: 193

　　In words: _____ hundred _____-_____

2.NBT.3

5) Use the base-ten blocks to write the three-digit number. Show the number in words.

　　Three digit number: _____

　　In words: _____

2.NBT.3

Do you understand?   ?   ✓

Affirmation: If I messed up yesterday, today is a new day, and I can try again.

Name: _____ Date: _____

Unit 5 Practice 7: I can represent three-digit numbers in different ways.

1) Show the three-digit number with base-ten blocks. Three-digit number: 264

2.NBT.3

2) Show the three-digit number **one hundred thirty-seven** with base-ten blocks.

Three-digit number: _____

2.NBT.3

3) Write the number shown in base-ten blocks in expanded form and in words.

Expanded form: _____

In words: _____

2.NBT.3

Do you understand?   ?

Affirmation: If I messed up yesterday, today is a new day, and I can try again.

Name: _____ Date: _____

Unit 5 Practice 7: I can represent three-digit numbers in different ways.

4) Show 5 hundreds, 1 ten, 8 ones with base-ten blocks, as a three-digit number and in expanded form.

Base-ten blocks:

Three-digit number: _____

Expanded form: _____

2.NBT.3

5) Finish the base-ten representation to show **four hundred five**. Write as a three-digit number and in expanded form.

Three-digit number: _____

Expanded form: _____

2.NBT.3

Do you understand?   ?   ✓

Affirmation: Math helps me develop grit and perseverance.

Name: _____ Date: _____

Unit 5 Practice 8: I can count tens and hundreds on a number line.

1) Count on by tens.

   0, 10, _____, _____, _____, _____, _____, _____, _____, _____, 100

   Write numbers to compete the number line.

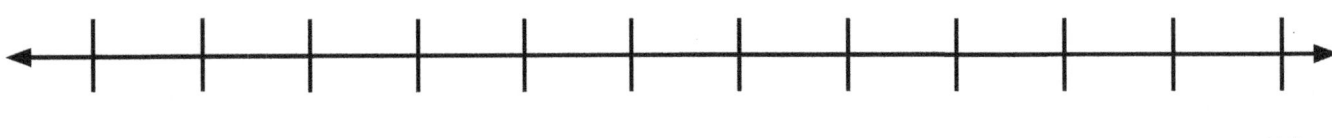

2.NBT.2

2) Count on by hundreds.

   0, 100, _____, _____, _____, _____, _____, _____, _____, _____, 1,000

   Write numbers to compete the number line.

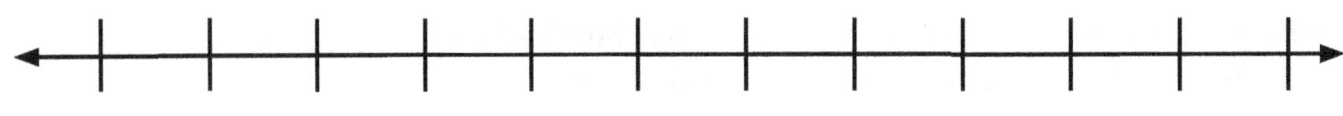

2.NBT.2

3) Draw and label a point to show where 850 is on the number line.

2.NBT.2

Do you understand?   ?   ✓

Name: _____ Date: _____

Unit 5 Practice 8: I can count tens and hundreds on a number line.

4) Draw and label a point to show where 90 would be on the number line.

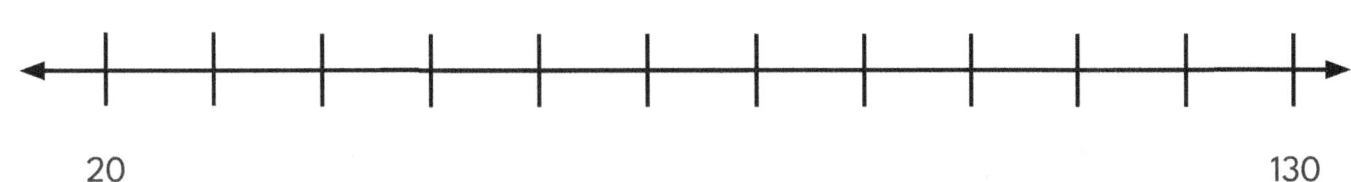

20   130

2.MD.6

5) Draw and label a point to show where 300 would be on the number line.

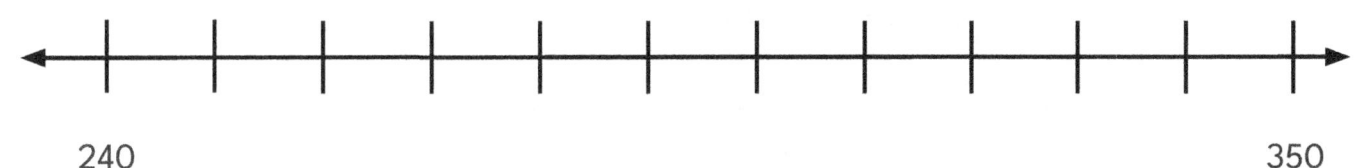

240   350

2.MD.6

6) Draw and label a point to show where 200 would be on the number line.

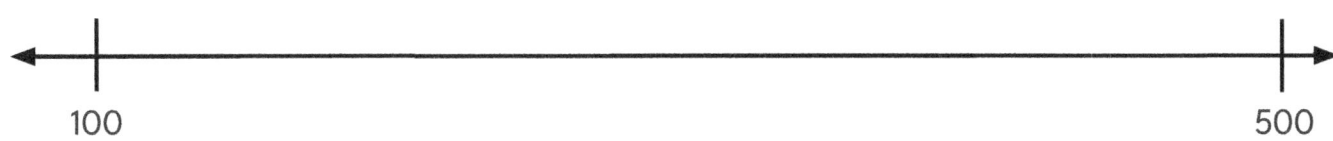

100   500

2.MD.6

Do you understand?   ?   ✓

I Can Do Math Practice Problems, Grade 2

Affirmation: I don't need to be fast at math. I need to understand and that takes time.

Name: _____   Date: _____

Unit 5 Practice 9: I can compare three-digit numbers.

1) Use the base-ten blocks to compare 147 and 253.
   Circle "is greater than" **or** is "less than".

147

is greater than

is less than

253

2.NBT.4

2) Use the number line to compare 420 and 360.
   Circle "is greater than" **or** "is less than".

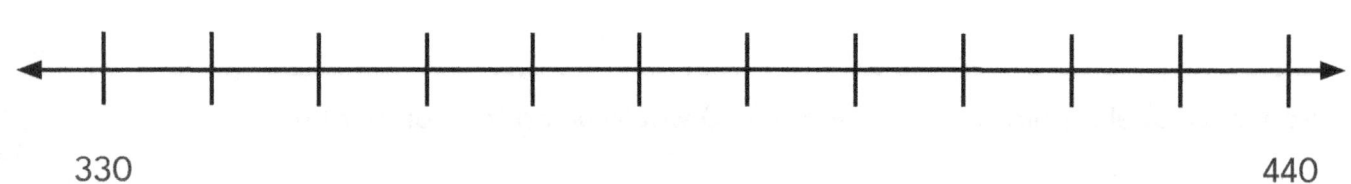

330   440

420   is greater than   is less than   360

2.NBT.4

Do you understand?

Affirmation: I don't need to be fast at math. I need to understand and that takes time.

Name: _____ Date: _____

Unit 5 Practice 9: I can compare three-digit numbers.

3) Draw base-ten blocks to model the numbers 425 and 452.
   Complete the number sentence using <, >, or =.

425 ◯ 452

2.NBT.4

4) Draw and label a point on the number line to show a number greater than 300.
   Draw and label a point on the number line to show a number less than 300.

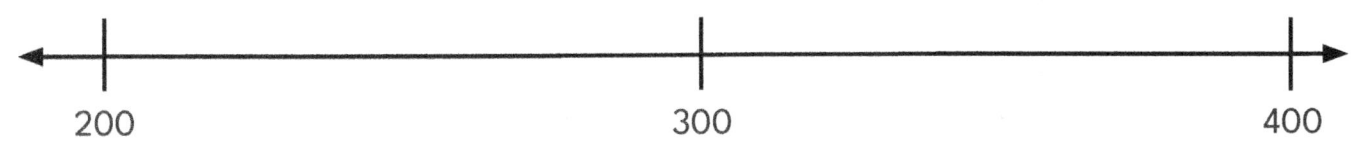

_____ > 300   _____ < 300

2.NBT.4

Do you understand? ? ✓

Affirmation: Think positive thoughts. Negative thoughts don't help us learn and grow.

Name: _____ Date: _____

Unit 5 Practice 10: I can use base-ten representations to compare numbers.

1) Add ten or subtract ten to solve.

    135 + 10 = _____   56 + 10 = _____   156 + 20 = _____

    149 + 20 = _____   48 - 10 = _____   167 - 20 = _____

    325 + 30 = _____   78 - 20 = _____   293 - 30 = _____

2.NBT.8

2) Add hundreds or subtract hundreds to solve.

    253 + 100 = _____   56 + 400 = _____   615 - 200 = _____

    952 - 200 = _____   478 - 100 = _____   799 - 500 = _____

    526 + 300 = _____   378 + 300 = _____   402 - 200 = _____

2.NBT.8

3) Count and compare the collections of base-ten blocks using <, >, or =.

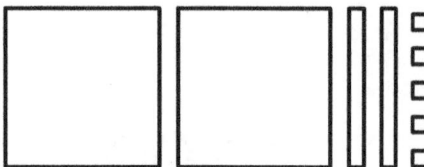 

_____ 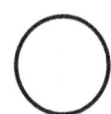 _____

2.NBT.4

Do you understand?   ?   ✓

Affirmation: Think positive thoughts. Negative thoughts don't help us learn and grow.

Name: _____ Date: _____

Unit 5 Practice 10: I can use base-ten representations to compare numbers.

4) Draw base-ten representations to compare 302 and 320.
   Complete the number sentence using <, >, or =.

$$302 \bigcirc 320$$

2.NBT.4

5) **Agree or Disagree**:
   Betty and Cameron each have a collection of five base-ten blocks. Betty has 2 hundreds 1 ten and 2 ones. Cameron has 2 hundreds, 2 tens and 1 one. Betty says that their collections are equal because they both have five base-ten blocks. Cameron says they both have five blocks but that his collection is greater. Who do you agree with?

   I agree with _____.

   Use drawings, numbers or words to show your thinking.

2.NBT.4

Do you understand?   ?   ✓

Affirmation: Mathematics makes me smarter because it makes me think!

Name: _____   Date: _____

Unit 5 Practice 11: I can use place value to compare three-digit numbers.

1) Match a number that makes each statement true.

200 + 5 + 10 = ☐   180

5 + 20 + 300 > ☐   660

900 - 240 = ☐   215

600 - 430 < ☐   253

2.NBT.4

2) Use the number line to complete each statement using <, >, or =.

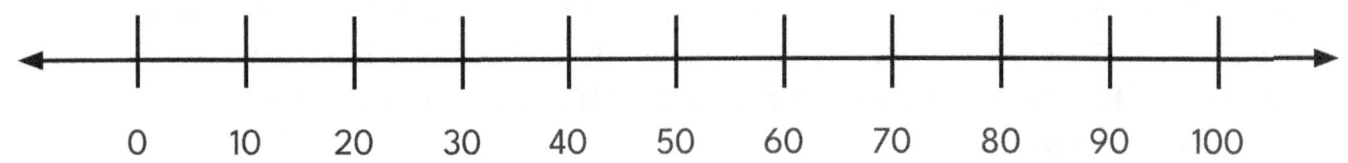

10 ◯ 60   47 ◯ 50   97 ◯ 79

20 ◯ 40   16 ◯ 9   33 ◯ 43

2.NBT.4

Do you understand?   ?   ✓

Affirmation: Mathematics makes me smarter because it makes me think!

Name: _____ Date: _____

Unit 5 Practice 11: I can use place value to compare three-digit numbers.

3) Haley wrote the place value riddle, "I have 4 ones, 6 tens and 3 hundreds." Blake wrote the place value riddle, "I have 4 hundreds, 3 tens and 6 ones."
Compare their numbers to decide who has the **lesser** number.

_____ has the **lesser** number.

2.NBT.4

4) **Agree or Disagree:**
Adrian wrote 832 = 823. He explained that the numbers are equal because they have the same digits 8, 2, and 3.
Do you agree or disagree with Adrian?

I _____ with Adrian.

2.NBT.4

Do you understand?   ?   ✓

Affirmation: I am human and humans are mathematical beings.

Name: _____ Date: _____

Unit 5 Practice 12: I can order numbers from least to greatest.

1) Circle true or false for each equation.

707 = 300 + 400 + 7

300 + 400 + 300 = 1,000

60 + 60 + 400 = 466

500 + 50 + 5 = 560

2.NBT.8

2) Place the numbers 70, 80, 40, 50, 20 on the number line.

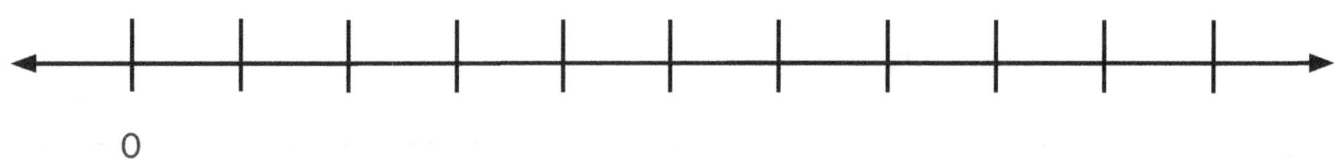

0

Order the numbers from **least to greatest**.

_____, _____, _____, _____, _____

2.NBT.4

3) Place the numbers 700, 800, 400, 500, 200 on the number line.

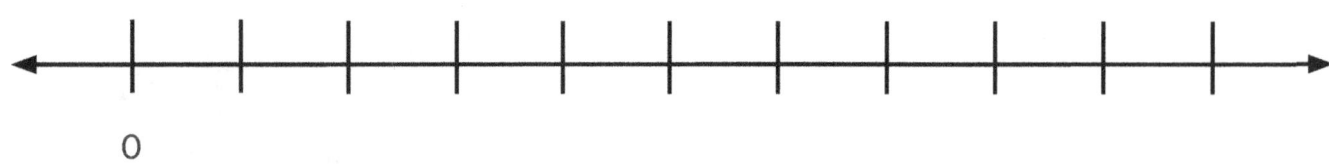

0

Order the numbers from **least to greatest**.

_____, _____, _____, _____, _____

2.NBT.4

Do you understand?

Affirmation: I am human and humans are mathematical beings.

Name: _____ Date: _____

Unit 5 Practice 12: I can order numbers from least to greatest.

4) Place the numbers 342, 328, 310, 346, and 330 on the number line.

Order the numbers from **least to greatest**.

_____, _____, _____, _____, _____

Explain how the numbers on the number line are the same or different from the numbers you ordered from least to greatest.

2.NBT.4

5) Write each as a three-digit number.

400 + 20 + 9 = _____   6 hundreds + 5 ones + 3 tens = _____

50 + 500 + 6 = _____   2 hundreds + 12 tens = _____

40 + 4 + 800 = _____   one hundred fifteen _____

Order the three-digit numbers from **least to greatest**.

_____, _____, _____, _____, _____, _____

2.NBT.1b

Do you understand?   ?   ✓

Affirmation: Even when we know the answer, it can be fun to try out a different way.

Name: _____   Date: _____

Unit 5 Practice 13: I can order numbers from greatest to least.

1) Count back by ones.

   97, _____, _____, _____, _____, _____, _____, _____, _____, 88

   Count back by tens.

   410, _____, _____, _____, _____, _____, _____, _____, 330

   Count back by hundreds.

   997, _____, _____, _____, _____, _____, _____, _____, 197

   2.NBT.2

2) Place the numbers 90, 20, 40, 60, 30 on the number line.

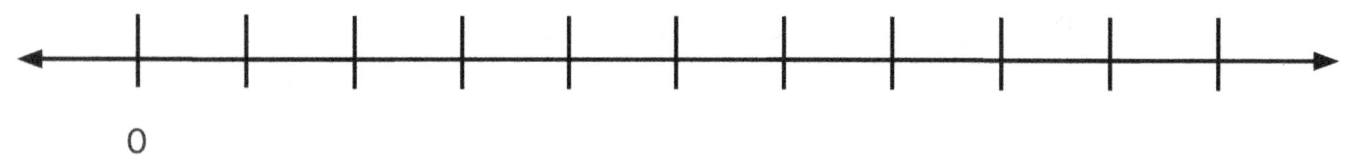

0

Order the numbers from **least to greatest**.

_____, _____, _____, _____, _____

Order the numbers from **greatest to least**.

_____, _____, _____, _____, _____

2.NBT.4

Do you understand?   ?

Affirmation: Even when we know the answer, it can be fun to try out a different way.

Name: _____ Date: _____

Unit 5 Practice 13: I can order numbers from greatest to least.

3) Place the numbers 524, 538, 550, 547, 542 on the number line.

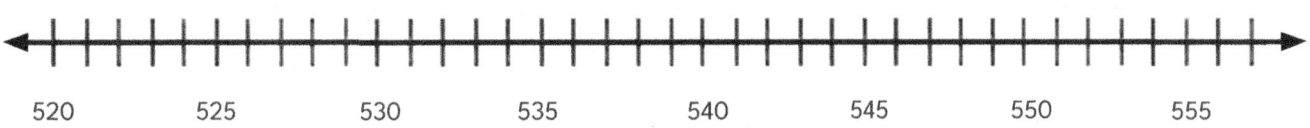

Order the numbers from **greatest to least**.

_____, _____, _____, _____, _____

Explain how the numbers on the number line are the same or different from the numbers you ordered from greatest to least.

2.NBT.4

4) Write each as a three-digit number.

90 + 4 + 400 = _____   six hundred thirty-seven _____

700 + 30 + 6 = _____   3 hundreds + 18 tens + 2 ones = _____

30 + 2 + 100 = _____   2 hundred + 8 ones + 4 tens = _____

Order the three-digit numbers from **greatest to least**.

_____, _____, _____, _____, _____, _____

2.NBT.1b

Do you understand?   ?   ✓

Affirmation: I have a powerful mind!

Name: _____ Date: _____

Unit 5 Practice 14: I can count collections to 1,000.

1) Count the base-ten block collection to find the total.

The total of this collection of base-ten blocks is _____.

2.NBT.2

2) MaryAnne has a collection of base-ten blocks that includes 7 hundreds, 21 tens and 20 ones. Is MaryAnne's collection of base-ten blocks greater than 1,000 or less than 1,000?

Choose: MaryAnne's collection [is greater than]   [is less than] 1,000.

2.NBT.4

Do you understand?   ?   ✓

Name: _____   Date: _____

Unit 5 Practice 14: I can count collections to 1,000.

3) Wendell wants to model 1,000 using base-ten blocks. He made this collection.

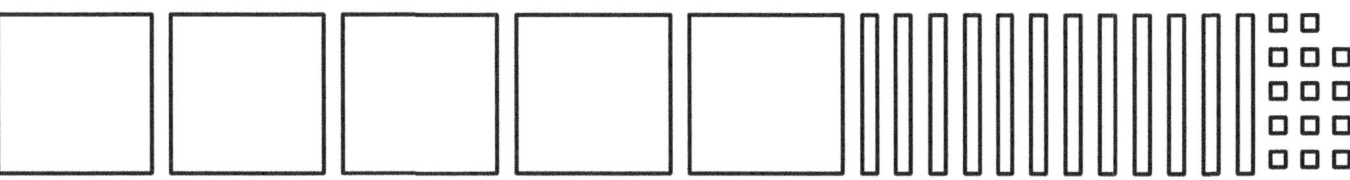

Draw a collection of base-ten blocks that would complete Wendell's model of 1,000.

Equation: _____ + _____ = 1,000

2.NBT.7a

4) Jaxon is using base-ten blocks to model the target number 758. He spilled out a bag of base-ten blocks modeled below.

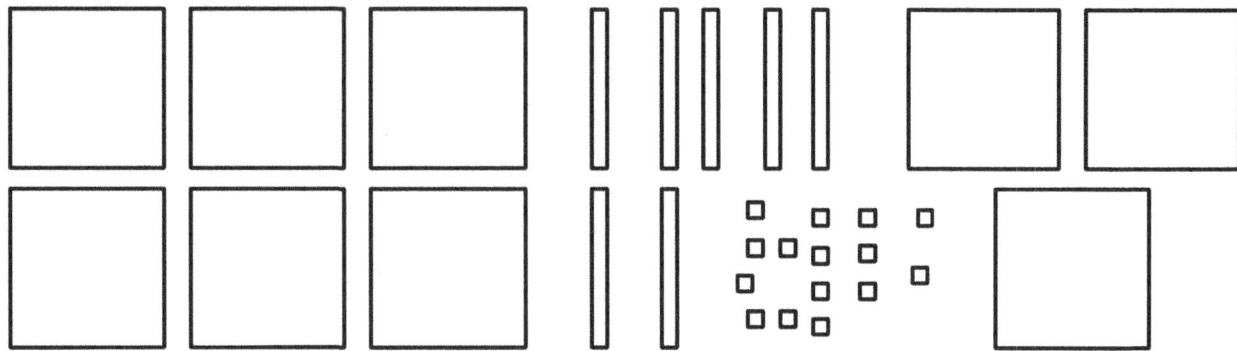

The collection is greater than his target number. Cross out base-ten blocks so that Jaxon's collection will equal the target number 758.

Equation: _____ - _____ = 758

2.NBT.7a

Do you understand?   ?   ✓

Affirmation: I get better the more I practice.

Name: _____ Date: _____

I can practice grade level fluencies.

Set 1: Add or subtract hundreds

1) 356 + 100 = _____

2) 715 + 100 = _____

3) 356 - 100 = _____

4) 715 - 100 = _____

5) 281 + 200 = _____

6) 281 - 200 = _____

7) 972 - 300 = _____

2.NBT.2

Set 2: Find the sum or difference

1) 245 + 14 = _____

2) 143 + 23 = _____

3) 245 - 14 = _____

4) 143 - 23 = _____

5) 369 + 110 = _____

6) 491 - 140 = _____

7) 972 - 320 = _____

2.NBT.5

Do you understand?   ?   ✓

Unit Reflection

Numbers to 1,000

Use this space to reflect on your understanding of the unit skills and concepts.

Skill/Concept	I can...	I need to work on...

Unit 6
Geometry, Time, and Money

In this unit, we will learn about different shapes and how to split them into equal parts. We will also tell time to the nearest five minutes and solve story problems using coins and dollars.

Two-Dimensional Shapes

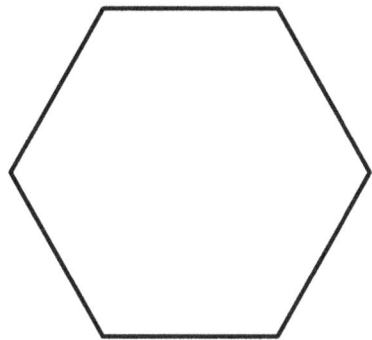

The hexagon has 6 sides and 6 angles

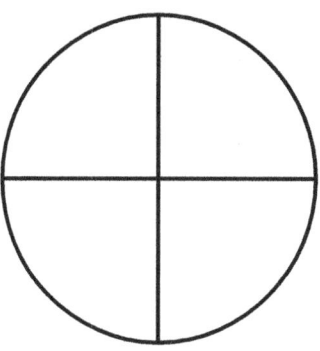

The circle is partitioned into 4 equal pieces

Analog Clock

The Clock Face Shows 9:10

Money

A Dollar Bill

Coins: penny, nickel, dime, quarter

Unit Vocabulary

Geometry, Time, and Money

Use this space to visualize the math vocabulary for this unit.

Word or Phrase	Example or Attributes	Visual Reminder

Unit Models & Strategies

Geometry, Time, and Money

Use this space to visualize the math models and strategies for this unit.

Model or Strategy	This is a...	It is used to...

Affirmation: I am brilliant, bright, and getting better every day!

Name: _____ Date: _____

Unit 6 Practice 1: I can name and list attributes of flat shapes.

1) Match each flat shape to its name.

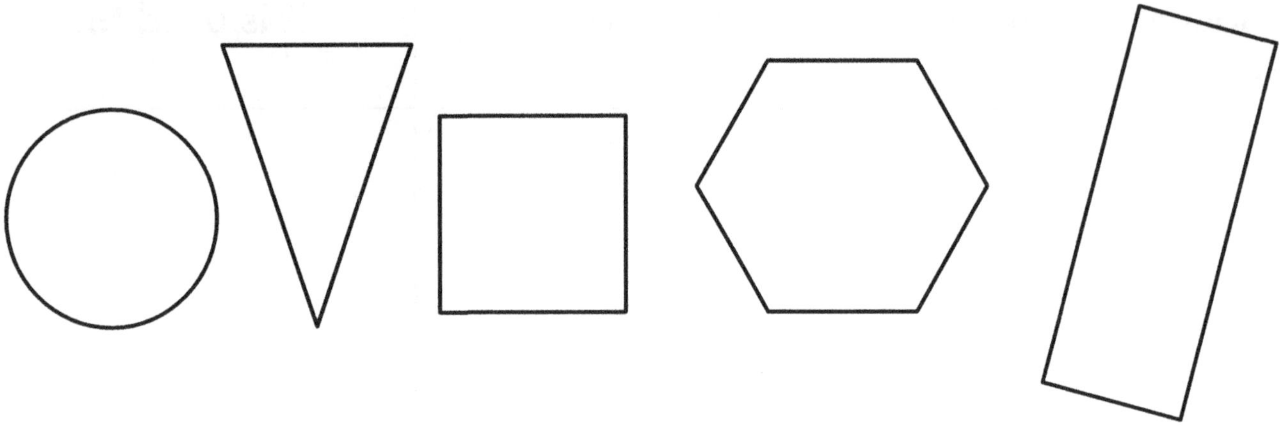

| square | circle | triangle | rectangle | hexagon |

1.G.1

2) Complete the chart for each shape.

Picture of shape	▭	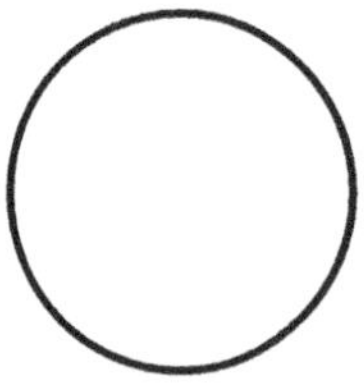 ◯	◻	 △
Name of shape				
Number of sides				
Number of corners				

1.G.2

Do you understand?   ?   ✓

Affirmation: I am brilliant, bright, and getting better every day!

Name: _____ Date: _____

Unit 6 Practice 1: I can name and list attributes of flat shapes.

3) Match each flat shape to its name. *Shapes may share a name.*

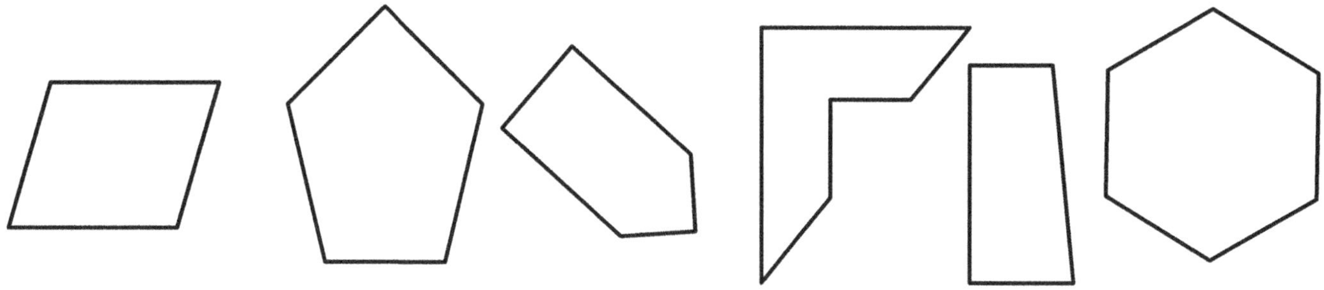

| pentagon | quadrilateral | hexagon |

4) Complete the chart for each shape.

Picture of shape	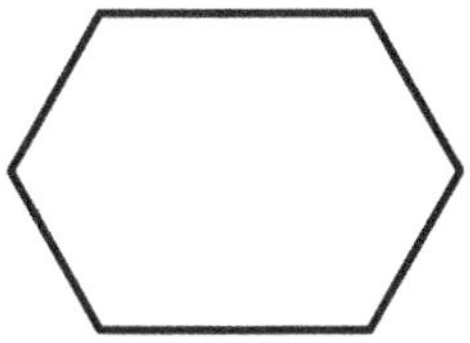	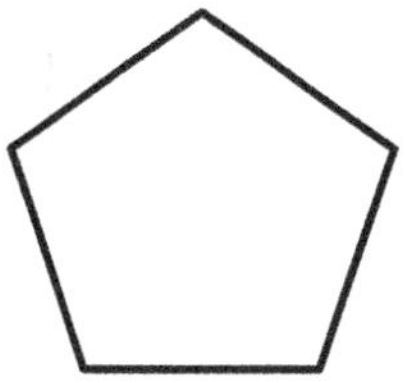	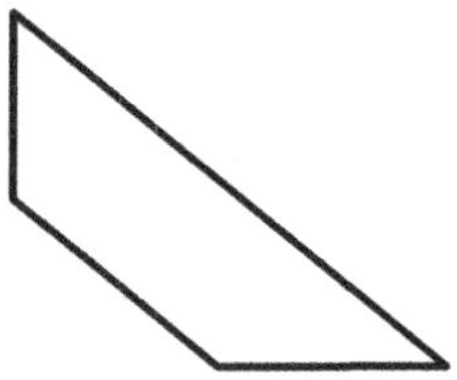
Name of shape			
Number of sides			
Number of corners			

Do you understand?   ?   ✓

Affirmation: Answers are important, but questioning and solving are just as important.

Name: _____ Date: _____

Unit 6 Practice 2: I can draw flat shapes.

1) Draw and label a **square**, a **triangle** and a **rectangle** on the dot grid.

1.G.1

2) Circle true 👍 or false 👎 for each named shape.

hexagon   quadrilateral   quadrilateral
👍 👎   👍 👎   👍 👎

pentagon   hexagon   pentagon
👍 👎   👍 👎   👍 👎

2.G.1

Do you understand?   ?   ✓

Name: _____   Date: _____

Unit 6 Practice 2: I can draw flat shapes.

3) Draw and label 3 different **quadrilaterals** on the dot grid.

2.G.1

4) Draw and label a **hexagon** and a **pentagon**.

2.G.1

Do you understand?   ?   ✓

Affirmation: New solutions happen when we try new ways of doing things.

Name: _____ Date: _____

Unit 6 Practice 3: I can use a 1 inch dot grid to draw shapes.

1) Draw a **square** with a side length of 3 inches.

2.MD.1

2) Draw and label a **rectangle** with a long side of 3 inches and a shorter side of 1 inch.

2.MD.1

Do you understand?   ?   ✓

Affirmation: New solutions happen when we try new ways of doing things.

Name: _____ Date: _____

Unit 6 Practice 3: I can use a 1 inch dot grid to draw shapes.

3) Clair drew some shapes on the 1 inch dot grid. Name each shape and label the sides of each shape that are 2 inches long.

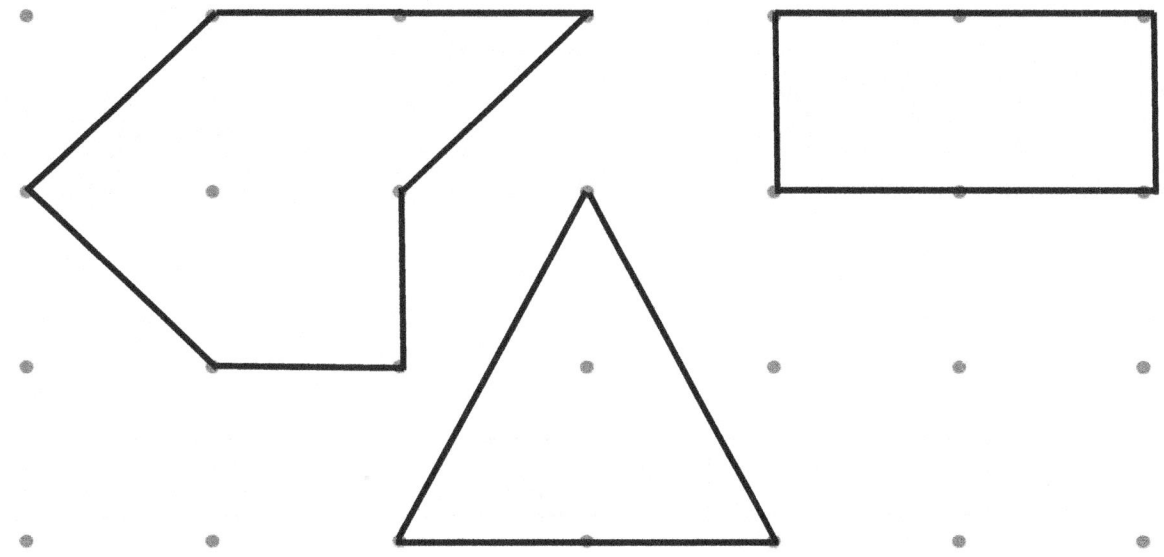

4) Finish the drawing to create a **pentagon**.

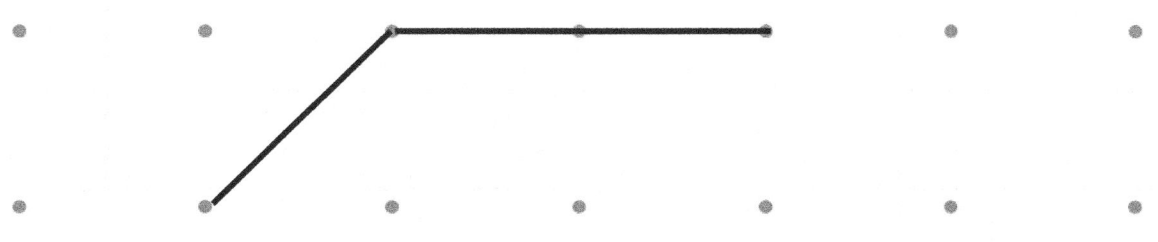

Do you understand?   ?   ✓

Affirmation: By asking questions, I learn more and help others learn as well.

Name: _____ Date: _____

Unit 6 Practice 4: I can name and describe solid shapes.

1) Match each solid shape to its name.

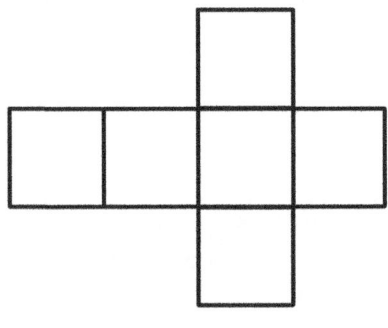

| cone | cylinder | sphere | cube |

1.G.1

2) Count and name the faces.

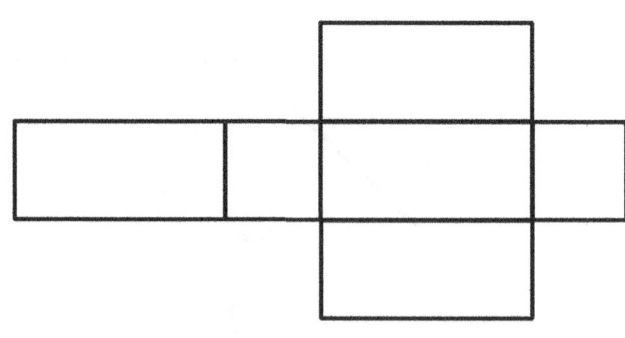

cube design   **rectangular prism design**

The cube has _____ faces.   The rectangular prism has _____ faces.

The faces of the cube are all   The faces of the rectangular prism are

_____.   _____ and _____.

2.G.1

Do you understand?

I Can Do Math Practice Problems, Grade 2

Affirmation: By asking questions, I learn more and help others learn as well.

Name: _____ Date: _____

Unit 6 Practice 4: I can name and describe solid shapes.

3) What solid shape will each shape design make?

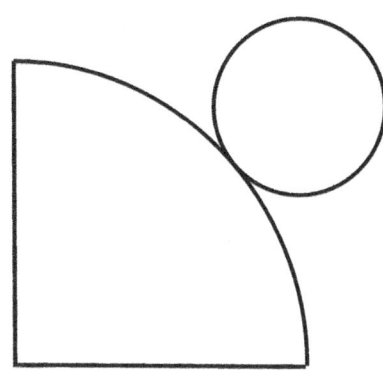   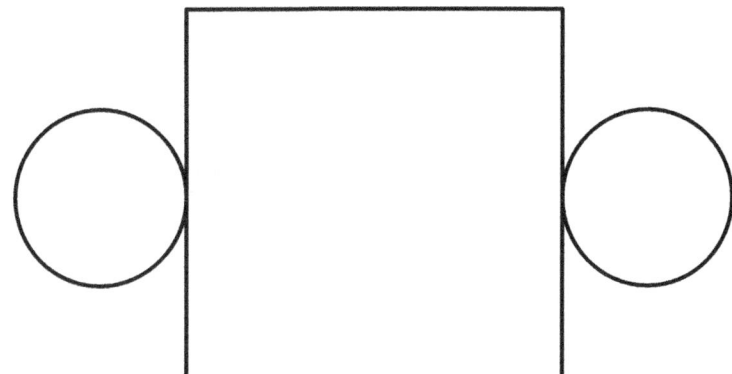

This shape design will make a _____.

This shape design will make a _____.

2.G.1

4) **Agree or Disagree**:
Jada drew this shape design to make a cube. Do you agree or disagree with Jada?

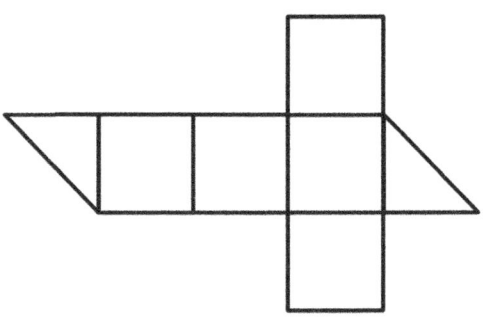

I _____ with Jada.

2.G.1

Do you understand?   ?   ✓

Affirmation: I love learning! I am a learner.

Name: _____   Date: _____

Unit 6 Practice 5: I can describe and compare flat shapes and solid shapes.

1) Find the sum.

    237 + 20 = _____   56 + 14 = _____   254 + 130 = _____

    542 + 40 = _____   83 + 20 = _____   712 + 34 = _____

    268 + 22 = _____   59 + 42 = _____   993 + 17 = _____

2.NBT.5

2) Tyler drew a flat shape that has 4 sides. None of the sides are the same length. Draw a shape that could be Tyler's shape.

Tyler's flat shape could be a _____.

2.G.1

3) Nadia drew this shape design to build a solid shape. Name the faces in Nadia's shape. What solid shape could Nadia make?

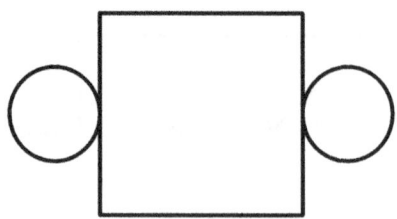

The faces in Nadia's shape design are _____ and _____.

Nadia's shape could be a _____.

2.G.1

Do you understand?

Affirmation: I love learning! I am a learner.

Name: _____ Date: _____

Unit 6 Practice 5: I can describe and compare flat shapes and solid shapes.

4) Write a shape riddle using attributes of a **hexagon**.

Draw the hexagon you are describing in your shape riddle.

2.G.1

5) Same or different. Which are the same? Which are different?

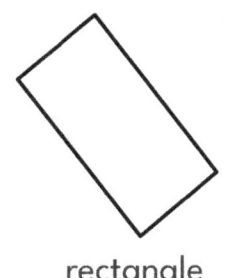
rectangle

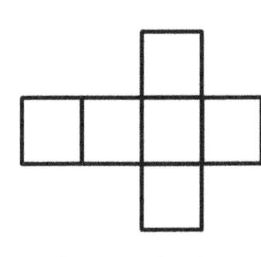
shape design

triangle

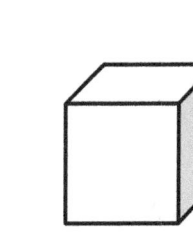
cube

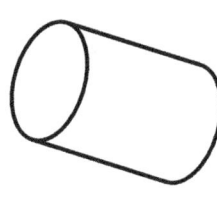
cylinder

Same	Different

2.G.1

Do you understand?   ?   ✓

Affirmation: Everyone can become better at math! Including ME!

Name: _____ Date: _____

Unit 6 Practice 6: I can make a flat shape using smaller shapes.

1) Match each pattern block to its name.

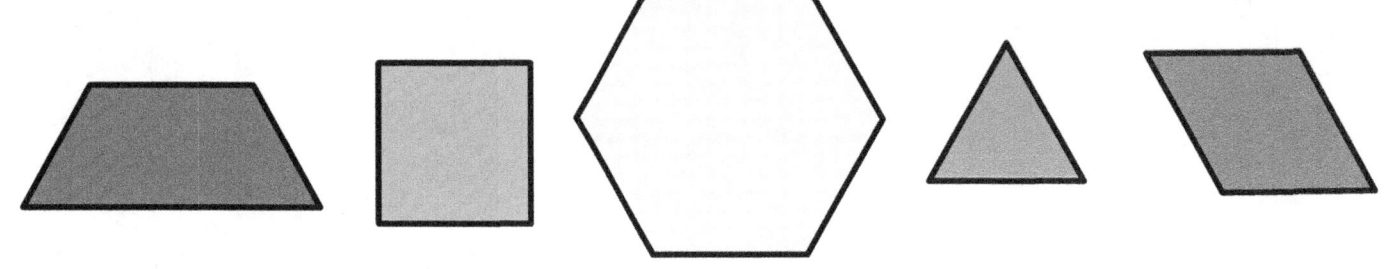

| hexagon | trapezoid | rhombus | square | triangle |

2) Omar made this design. How many of each pattern block shape did he use?

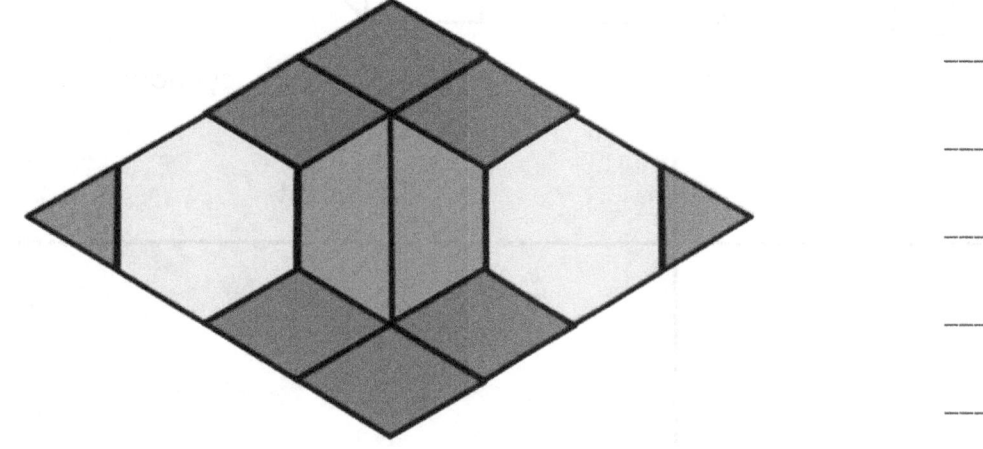

_____ hexagons

_____ trapezoids

_____ rhombuses

_____ squares

_____ triangles

Do you understand?   ?   ✓

Affirmation: Everyone can become better at math! Including ME!

Name: _____   Date: _____

Unit 6 Practice 6: I can make a flat shape using smaller shapes.

3) Draw to show how you can cover each hexagon with a smaller shape.

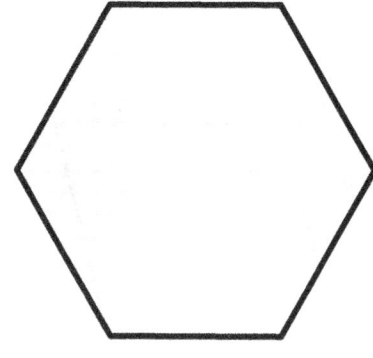 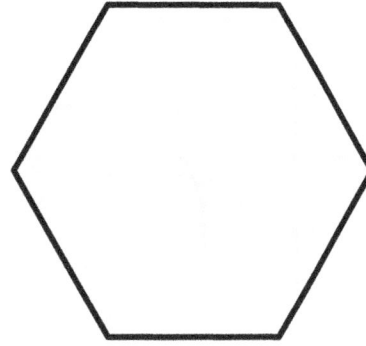 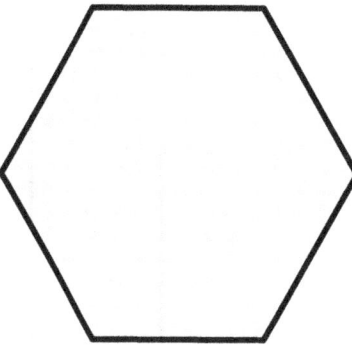

Cover with triangles.   Cover with rhombuses.   Cover with trapezoids.

I used _____ triangles.   I used _____ rhombuses.   I used _____ trapezoids.

4) Bryce and Penny used squares to make a new shape. How are their shapes the same? How are their shapes different?

Bryce's Shape    Penny's Shape

Same	Different

Do you understand?   ?   ✓

Affirmation: I can do harder math problems by drawing out the story.

Name: _____ Date: _____

Unit 6 Practice 7: I can partition shapes into equal pieces.

1) Name each shape. Partition each shape into 2 equal pieces.

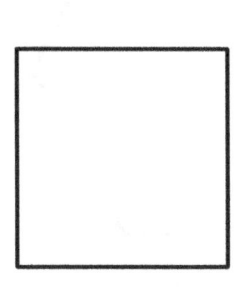  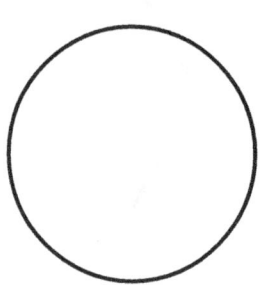 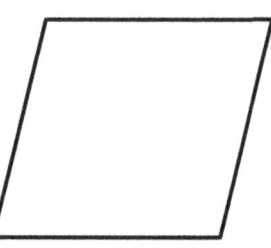

_____   _____   _____   _____

1.G.3

2) Circle true 👍 or false 👎. Each square is partitioned into equal size pieces.

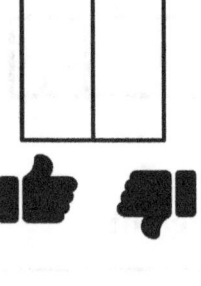 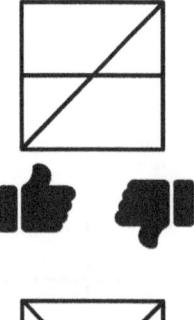 

👍 👎   👍 👎   👍 👎

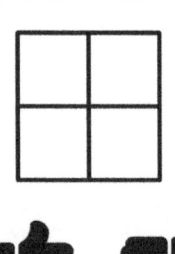 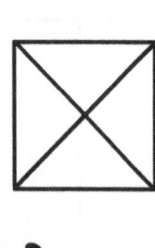 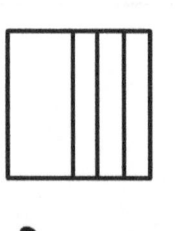

👍 👎   👍 👎   👍 👎

1.G.3

Do you understand?   ?   ✓

Affirmation: I can do harder math problems by drawing out the story.

Name: _____   Date: _____

Unit 6 Practice 7: I can partition shapes into equal pieces.

3) **Agree or Disagree:**
Zuri showed different ways she could partition a circle into fourths. Do you agree or disagree with Zuri?

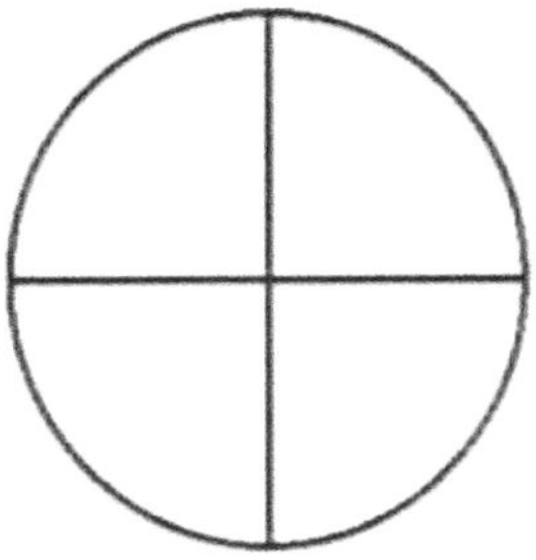

Way 1 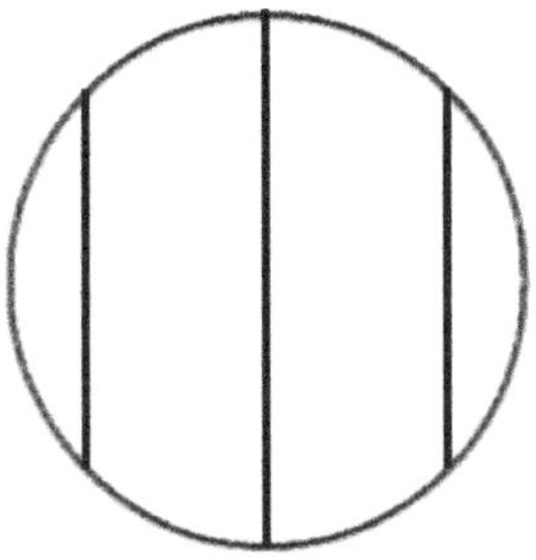   Way 2 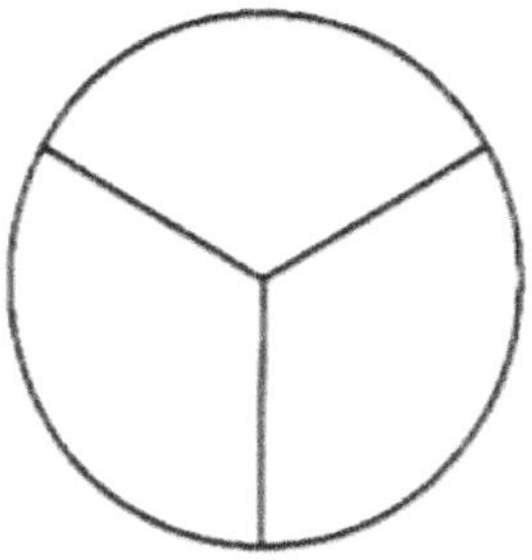   Way 3

I _____ with Zuri.

2.G.3

4) Partition each shape into **quarters**.

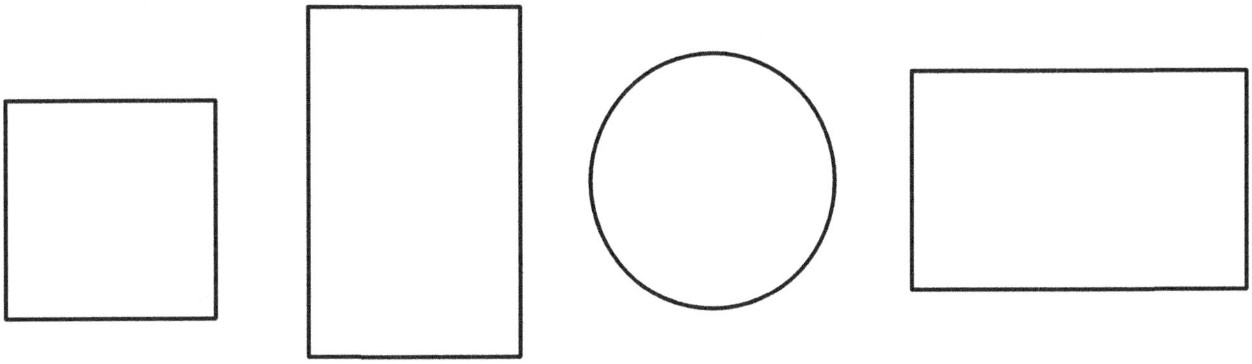

2.G.3

Do you understand?   ?   ✓

Affirmation: The more I talk about my thinking, the more connections I can make.

Name: _____ Date: _____

Unit 6 Practice 8: I can partition shapes into halves, thirds and fourths.

1) Find the difference.

       30 - 10 = _____   60 - 10 = _____   35 - 15 = _____

       30 - 5 = _____   60 - 20 = _____   45 - 15 = _____

       40 - 10 = _____   60 - 15 = _____   70 - 15 = _____

       40 - 5 = _____   60 - 5 = _____   70 - 55 = _____

2.NBT.2

2) Partition the squares into halves, thirds, or fourths. Shade one piece of each. How much of each square is shaded?

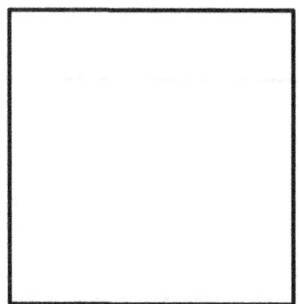   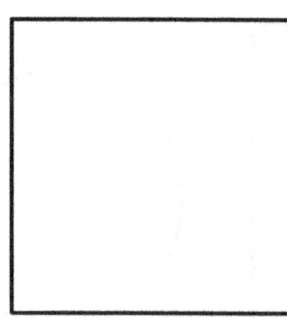   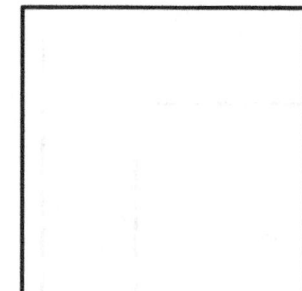

    show halves   show thirds   show fourths

_____ is shaded   _____ is shaded   _____ is shaded

2.G.3

Do you understand?   ?   ✓

Affirmation: The more I talk about my thinking, the more connections I can make.

Name: _____ Date: _____

Unit 6 Practice 8: I can partition shapes into halves, thirds and fourths.

3) Name each shape. Partition each shape into thirds.

_____   _____   _____

2.G.3

4) **Agree or Disagree**:
Skyler partitioned a square brownie into 4 pieces. He told 3 friends that if they share his brownie, each friend will get an equal size share. Do you agree or disagree with Skyler?

I _____ with Skyler.

2.G.3

Do you understand?   ?   ✓

Affirmation: Everyone, I mean everyone makes mistakes, especially in math.

Name: _____   Date: _____

Unit 6 Practice 9: I can show equal shares using different shapes.

1) Match to find the sum.

15 + 15 = ☐		65
40 + 15 = ☐		30
50 + 15 = ☐		40
15 + 25 = ☐		55

45 + 10 = ☐		60
15 + 20 = ☐		55
15 + 45 = ☐		75
60 + 15 = ☐		35

2.NBT.2

2) Partition the square into four equal pieces. Shade one of the pieces.
   How much of the square is shaded?
   How much of the square is **NOT** shaded?

_____ of the square is shaded.

_____ of the square is not shaded.

2.G.3

Do you understand?   ?   ✓

Affirmation: Everyone, I mean everyone makes mistakes, especially in math.

Name: _____ Date: _____

Unit 6 Practice 9: I can show equal shares using different shapes.

3) Maggie and her sister have two different types of cookies to share. Show how they can share the cookies equally.

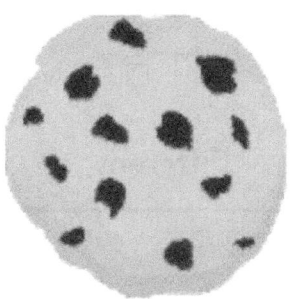 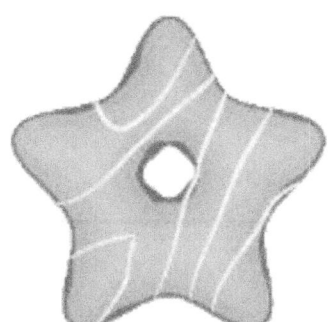

2.G.3

4) Show how three friends can share the cracker equally.

The cracker is partitioned into _____ equal pieces.

Each friend will get _____ of the cracker.

2.G.3

Do you understand?

Affirmation: I am smart in my own way and so is everyone else.

Name: _____ Date: _____

Unit 6 Practice 10: I can work with equal shares.

1) Elsa and Sawyer each have a collection of counters. How are their collections the same? How are their collections different?

Elsa's Counters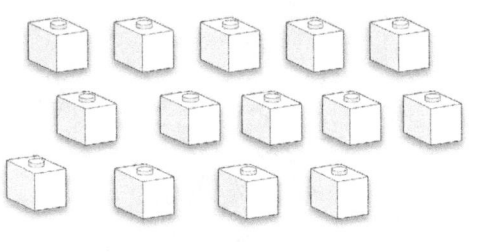

Sawyer's Counters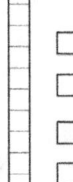

Same	Different

2.NBT.4

2) Naya partitioned a candy bar into equal size pieces. Naya wants to eat the whole candy bar. How many pieces will she eat?

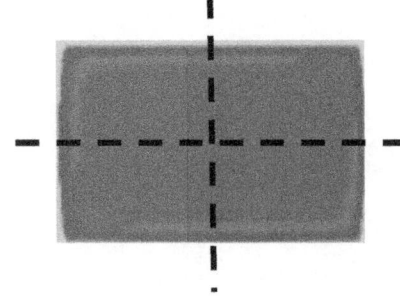

2.G.3

Do you understand?   ?   ✓

Affirmation: I am smart in my own way and so is everyone else.

Name: _____ Date: _____

Unit 6 Practice 10: I can work with equal shares.

3) Zuri is making a new shape with pattern blocks. Her new shape will show fourths. Finish Zuri's shape.

2.G.3

4) Destiny partitioned a cookie into quarters. She ate 3 of the pieces. How much of the cookie did she eat?

Destiny ate _____ of the cookie.

How much of the cookie is left?

There is _____ of the cookie left.

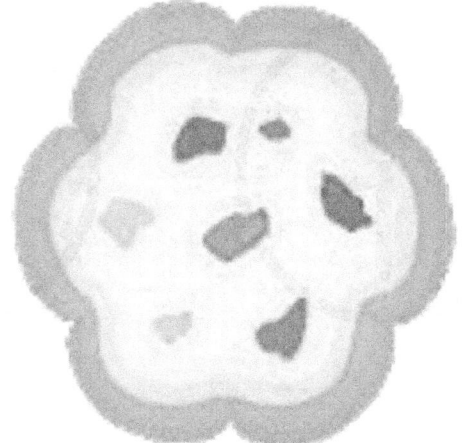

2.G.3

Do you understand?

Affirmation: I can use anything as a math manipulative.

Name: _____   Date: _____

Unit 6 Practice 11: I can use halves and quarters to tell time on an analog clock.

1) Shade to show the fraction.

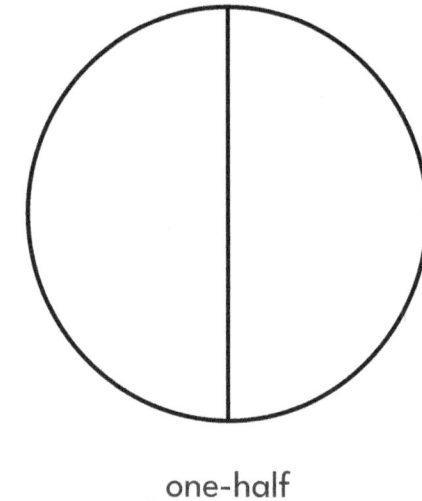   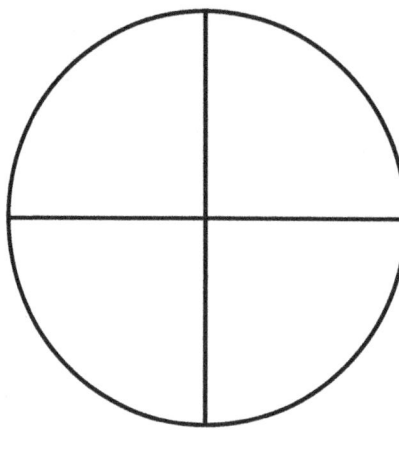   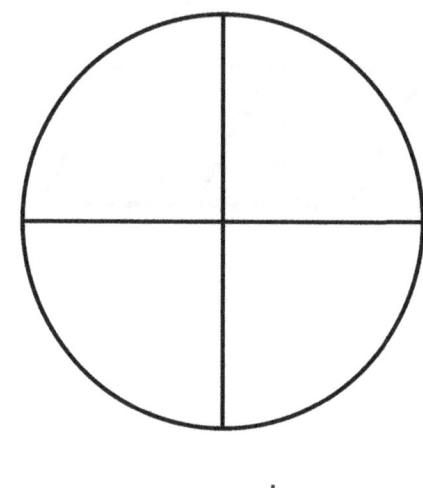

one-half   one-fourth   one-quarter

1.G.3

2) Cross out the clock that does **not** show **half past**.

1.G.3

Do you understand?   ?   ✓

Affirmation: I can use anything as a math manipulative.

Name: _____ Date: _____

Unit 6 Practice 11: I can use halves and quarters to tell time on an analog clock.

3) Draw hands to show the time on the analog clock.

 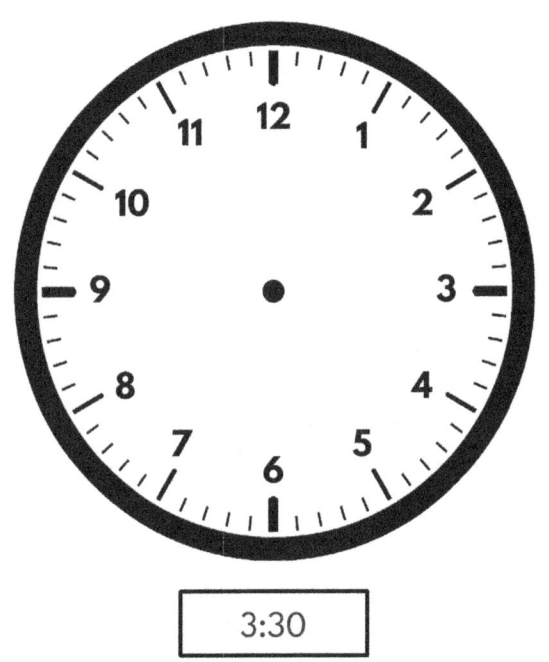

3:00   3:30

1.MD.3

4) Draw hands to show the time on the analog clock.
   Explain how you used the minute hand to show each time.

quarter past 4   quarter to 5

2.MD.7

Do you understand?   ?   ✓

Affirmation: I don't give up!

Name: _____   Date: _____

Unit 6 Practice 12: I can tell time using an analog clock.

1) Count on by fives.

   5, _____, _____, _____, _____, _____, _____, _____, _____, 50

   15, _____, _____, _____, _____, _____, _____, _____, _____, 60

   30, _____, _____, _____, _____, _____, _____, _____, _____, 75

   2.NBT.2

2) Place the missing numbers on the number line. Draw points on the number line to show 15, 30, 45, 60.

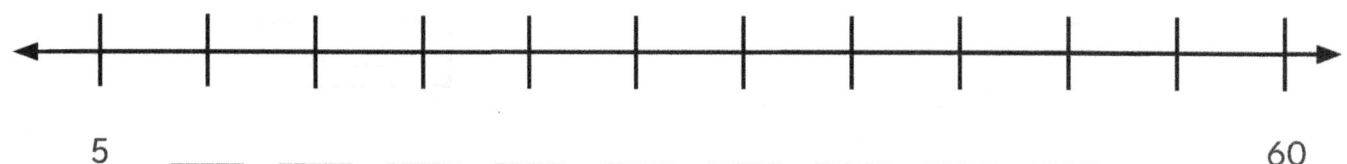

2.NBT.2

3) Which two numbers is the arrow between?

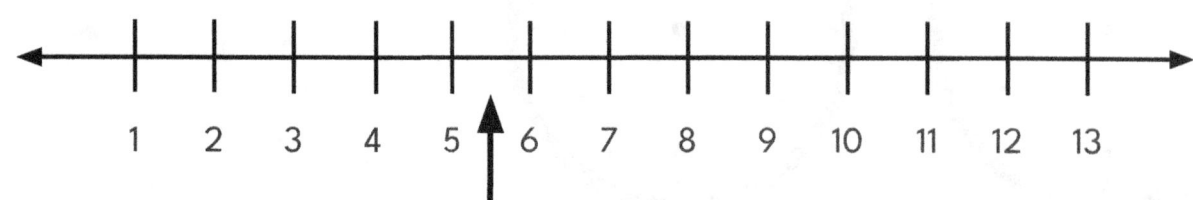

The arrow is pointing between the number _____ and the number _____.

K.G.1

Do you understand?   ?   ✓

Affirmation: I don't give up!

Name: _____ Date: _____

Unit 6 Practice 12: I can tell time using an analog clock.

4) Write the 5-minute intervals outside the analog clock.

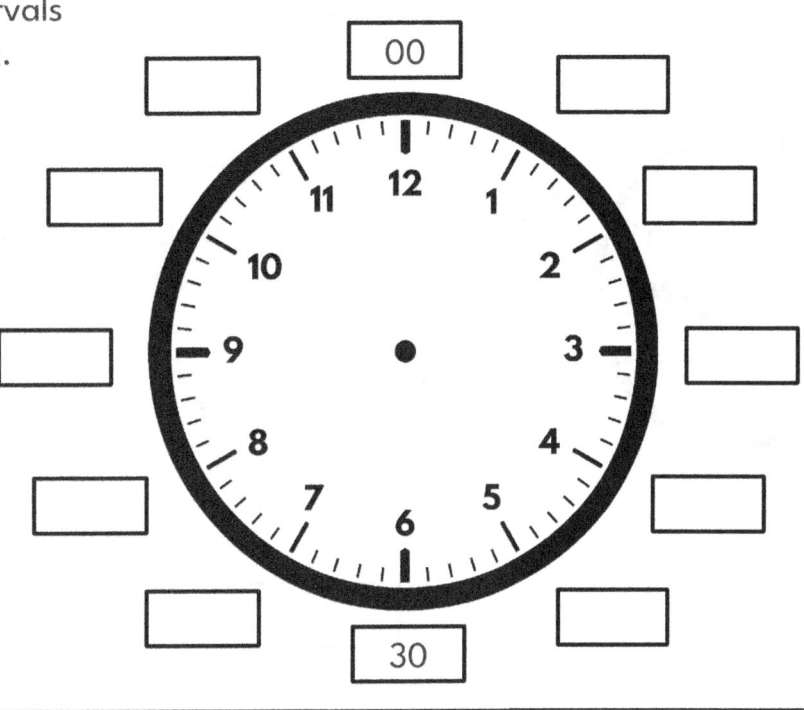

2.MD.7

5) Draw hands to show the time on the analog clock.

5 minutes past 6

10 minutes past 6

15 minutes past 6

2.MD.7

Do you understand?   ?   ✓

Affirmation: My brain never stops growing with ideas.

Name: _____   Date: _____

Unit 6 Practice 13: I can make a schedule to show a.m. or p.m.

1) Write the time using the words **half past**, **quarter past**, *or* **quarter till**. Show the time on the digital clock.

_____   _____   _____

      :   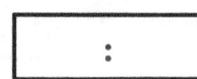   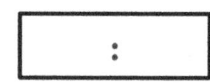

2.MD.7

2) **Agree or Disagree**:
Alex looked at the classroom analog clock and saw the time showed half past one. Alex said that it was **1:30 a.m.** because the sun was out. Do you agree or disagree with Alex?

I _____ with Alex.

2.MD.7

Do you understand?   ?   ✓

Affirmation: My brain never stops growing with ideas.

Name: _____ Date: _____

Unit 6 Practice 13: I can make a schedule to show a.m. or p.m.

3) Amir went to the State County Fair with his family. Amir made a schedule for the day. Complete the schedule that Amir might have made by filling in the activity. Tell if the activity took place in the a.m. or p.m.

Time	Activity	a.m or p.m.
9:00		
10:15		
11:45		
1:30		
5:30		
8:20		

Amir's activities for the State County Fair:

- ❏ ate dinner
- ❏ pet the horses
- ❏ watched fireworks
- ❏ bought admission tickets
- ❏ ate lunch
- ❏ rode the roller coaster

2.MD.7

Do you understand?   ?   ✓

Affirmation: Math helps me build confidence in my abilities.

Name: _____ Date: _____

Unit 6 Practice 14: I can show time on an analog clock and on a digital clock.

1) Find the sum.

15 + 15 = _____   50 + 5 + 5 = _____

15 + 30 = _____   20 + 20 + 5 = _____

15 + 15 + 15 = _____   30 + 15 + 15 = _____

10 + 10 + 20 + 5 = _____   5 + 5 + 10 + 10 = _____

30 + 15 + 10 + 5 = _____   20 + 15 + 5 + 5 = _____

2.NBT.5

2) Complete the analog clock.
   Draw hands to show 4 o'clock.

.MD.7

Do you understand?   ?   ✓

Affirmation: Math helps me build confidence in my abilities.

Name: _____ Date: _____

Unit 6 Practice 14: I can show time on an analog clock and on a digital clock.

3) Draw hands to show the time on the analog clock.

| quarter past 7 | half past 2 | 25 minutes past 8 |

2.MD.7

4) Show the time on a digital clock.

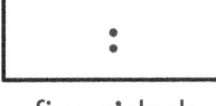 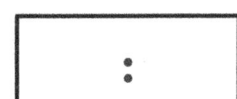 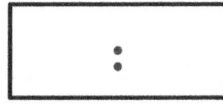

five o'clock   half past 4   quarter to six

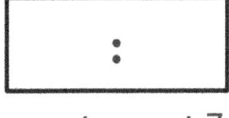 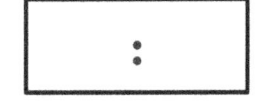 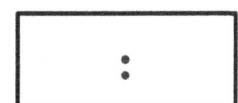

quarter past 7   10 minutes past 11   15 minutes to eight

2.MD.7

Do you understand?   ?   ✓

Affirmation: Every day that I keep trying, is one day closer to me getting it.

Name: _____ Date: _____

Unit 6 Practice 15: I can identify coin names and coin values.

1) Count on by tens.

   10, _____, _____, _____, _____, _____, _____, _____, _____, 100

   Count on by fives.

   5, _____, _____, _____, _____, _____, _____, _____, _____, 50

2.NBT.2

2) Match the coin to its name.

| penny | nickel | dime |

2.MD.8

3) Match the coin to its **value**.

| 10 cents | 5 cents | 1 cent |

2.MD.8

Do you understand?   ?   ✓

Affirmation: Every day that I keep trying, is one day closer to me getting it.

Name: _____ Date: _____

Unit 6 Practice 15: I can identify coin names and coin values.

4) What is the name of the coin in this collection?

There are _____ in this collection.

The **value** of this collection is _____.

2.MD.8

5) What is the name of the coin in this collection?

There are _____ in this collection.

The **value** of this collection is _____.

2.MD.8

6) What is the **value** of this mixed coin collection?

The **value** of this collection is _____.

2.MD.8

Do you understand?

I Can Do Math Practice Problems, Grade 2

Affirmation: Believing that I can be successful in math means that I can do it!

Name: _____ Date: _____

Unit 6 Practice 16: I can compare the value of coin collections.

1) Write the name of each coin and its **value**.

Name: _____   Name: _____   Name: _____   Name: _____

Value: _____   Value: _____   Value: _____   Value: _____

2.MD.8

2) Circle true 👍 or false 👎. The **value** of each coin collection is 25 cents..

👍 👎   👍 👎

👍 👎   👍 👎

2.MD.8

Do you understand?   ?   ✓

I Can Do Math Practice Problems, Grade 2

Affirmation: Believing that I can be successful in math means that I can do it!

Name: _____ Date: _____

Unit 6 Practice 16: I can compare the value of coin collections.

3) Bella and Elijah each have a collection of four coins. Find the value of each coin collection.

Bella's coin collection.
Bella's collection value: _____.

Elijah's coin collection.
Elijah's collection value: _____.

Which four-coin collection has the **greater value**?

_____ has the greater value.

2.MD.8

4) Jasmine is buying an eraser that costs 30 cents. Show two different coin combinations that have a value of 30 cents.

2.MD.8

Do you understand?

Affirmation: If I get lost, I will ask for help.

Name: _____   Date: _____

Unit 6 Practice 17: I can find coin combinations that have the value of one dollar.

1) Match to find the number that makes each equation true.

50 + 25 + ☐ = 100   20

25 + 10 + 10 + 5 + ☐ = 100   25

40 + 5 + 5 + 15 + ☐ = 100   50

☐ + 75 + 5 = 100   35

2.NBT.6

2) How many? How much?

How many coins in this collection?

There are _____ coins in this collection.

What is the **value** of this coin collection in cents?

The value is _____ cents.

2.MD.8

Do you understand?   ?   ✓

Affirmation: If I get lost, I will ask for help.

Name: _____ Date: _____

Unit 6 Practice 17: I can find coin combinations that have the value of one dollar.

3) Show two ways you can make one dollar using only coin combinations.

Way 1   Way 2

2.MD.8

4) Malik has this coin collection. Does he have more than a dollar or less than a dollar?

Malik has _____ than a dollar.

2.MD.8

Do you understand?

Affirmation: Solving problems is what humans do!

Name: _____   Date: _____

Unit 6 Practice 18: I can solve word problems using money.

1) Find the difference.

   100 - 75 = _____   75 - 20 = _____

   75 - 10 = _____   55 - 25 = _____

   80 - 30 = _____   100 - 45 = _____

   95 - 25 = _____   50 - 27 = _____

   100 - 55 = _____   90 - 15 = _____

   2.NBT.5

2) How **many**? How **much**?

How many coins in this collection?

There are _____ coins in this collection.

What is the **value** of this coin collection in cents?

The value is _____ cents.

2.MD.8

Do you understand?   ?

Affirmation: Solving problems is what humans do!

Name: _____ Date: _____

Unit 6 Practice 18: I can solve word problems using money.

Questions A and B refer to the information below.

School Store	
Snack	**Cost**
lollipop	10 ¢
bag of gumdrops	20 ¢
bag of chips	40 ¢
two cookies	25 ¢

A) Oden bought one lollipop and two cookies. How much did Oden spend at the school store on snacks?

Oden spent _____ on snacks.

2.MD.8

B) Luca has one dollar to spend at the school store on snacks. He wants to buy a bag of chips and four cookies. Does Luca have enough money?

Luca _____ have enough money.
   (does / does not)

2.MD.8

Do you understand?   ?   ✓

Affirmation: Not understanding something now, doesn't mean I won't in the future.

Name: _____   Date: _____

Unit 6 Practice 19: I can solve story problems using units of a dollar.

1) Add or subtract.

   36 + 4 = _____   73 - 3 = _____

   36 + 10 = _____   73 - 10 = _____

   36 + 14 = _____   73 - 13 = _____

   36 + 15 = _____   73 - 15 = _____

   36 + 25 = _____   73 - 25 = _____

   2.NBT.5

2) Phoebe had $19 in her piggy bank. She puts the money she received as a birthday gift into her piggy bank. She now has $42 in her piggy bank. How much money did Phoebe receive as a birthday gift?

   Phoebe received _____ as a birthday gift.

   2.OA.1

Do you understand?   ?   ✓

Affirmation: Not understanding something now, doesn't mean I won't in the future.

Name: _____ Date: _____

Unit 6 Practice 19: I can solve story problems using units of a dollar.

Questions A and B refer to the information below.

Pet Supplies	
Item	Cost
bag of dog food	$34
bag of cat food	$28
pack of dog treats	$12
pet bowl set	$19

A) Annalise has $50 to spend on pet supplies. She bought a bag of dog food and a pack of dog treats. How much money will she have left after paying for the two items?

Annalise will have _____ left.

2.OA.1

B) Jaden has $50. He wants to buy a bag of cat food and a pet bowl set. Does Jaden have enough money to buy the two items?

Jaden _____ have enough money.
     (does / does not)

2.OA.1

Do you understand?   ?   ✓

Name: _____ Date: _____

Unit 6 Practice 20: I can solve and write story problems using money.

1) Circle true 👍 or false 👎 for each number statement.

25 + 15 < 30 + 5

100 - 35 = 60 + 5

54 + 6 > 50 + 10

46 - 20 < 20 + 10

2.NBT.4

2) Naomi has a coin collection with 2 dimes, 3 nickels and 1 quarter. Chandler has a coin collection with 2 nickels, 3 dimes and 1 quarter.

Who has **more coins** in their collection?
Which coin collection has the **lesser value**?

2.MD.8

Do you understand?   ?   ✓

Affirmation: I love thinking! I am a thinker.

Name: _____ Date: _____

Unit 6 Practice 20: I can solve and write story problems using money.

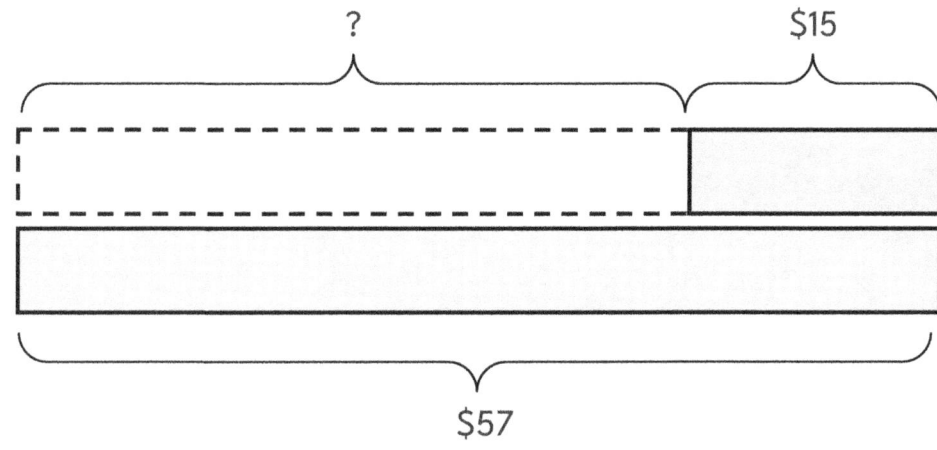

A) Write a story problem that matches the tape diagram.

B) Solve your story problem.

Equation: _____

Do you understand? ? ✓

Affirmation: When I am nervous or scared trying is brave and courageous!

Name: _____ Date: _____

Unit 6 Practice 21: I can make designs with pattern blocks.

1) Use the grid below to draw and label a hexagon, a triangle, a rhombus and a trapezoid from a pattern block set.

2) Aubrey is drawing a pattern block design using 1 hexagon, 3 trapezoids, and 2 triangles. She has drawn the hexagon and one of the trapezoids. Finish Aubrey's pattern block design.

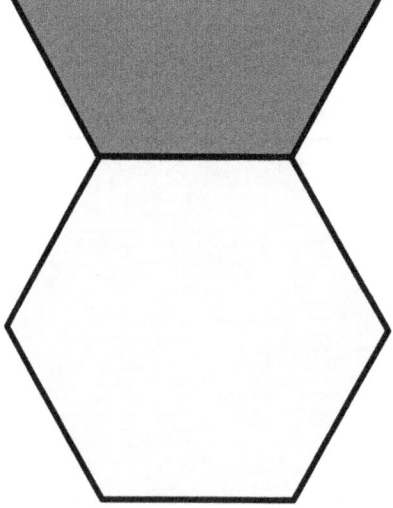

Do you understand?   ?   ✓

Affirmation: When I am nervous or scared trying is brave and courageous!

Name: _____ Date: _____

Unit 6 Practice 21: I can make designs with pattern blocks.

Use hexagons, rhombuses, trapezoids and triangles to make a pattern block design on the grid.

A) How many of each shape did you use in your design?

2.G.1

B) Write a pattern block puzzle for your design.

2.G.1

Do you understand?   ?   ✓

Affirmation: I get better the more I practice.

Name: _____ Date: _____

I can practice grade level fluencies.

Set 1: Find the sum

1) 20 + 20 = _____

2) 25 + 25 = _____

3) 30 + 15 = _____

4) 45 + 10 = _____

5) 25 + 24 = _____

6) 15 + 15 = _____

7) 10 + 25 = _____

2.NBT.2

Set 2: Find the difference

1) 25 - 10 = _____

2) 25 - 15 = _____

3) 25 - 20 = _____

4) 60 - 10 = _____

5) 60 - 15 = _____

6) 45 - 10 = _____

7) 45 - 15 = _____

2.NBT.2

Do you understand?   ?   ✓

Unit Reflection

Geometry, Time, and Money

Use this space to reflect on your understanding of the unit skills and concepts.

Skill/Concept	I can...	I need to work on...

Unit 7
Adding and Subtracting within 1,000

We will learn ways to add and subtract numbers up to 1,000 using what we know about place value and how addition and subtraction are connected. Sometimes we will break apart or put together tens and hundreds to help us solve problems.

Base-Ten Numerals
The number 329 has
3 hundreds, 2 tens, 9 ones

Addition and Subtraction
245 + 123 = 368
674 - 231 = 443

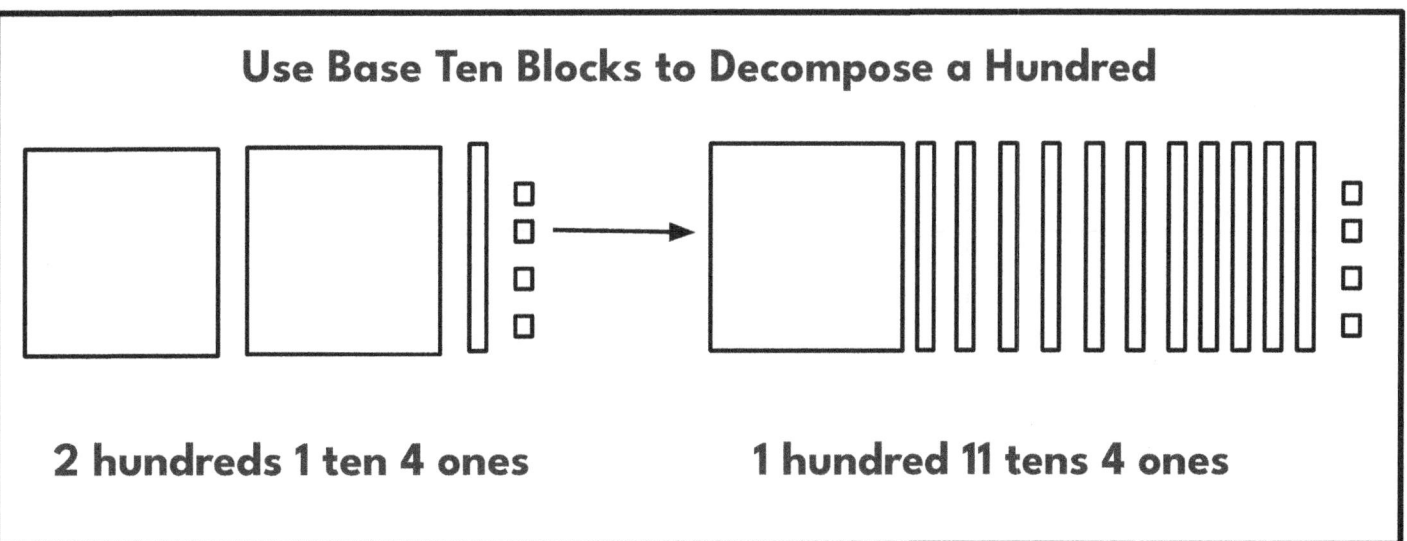

Use Base Ten Blocks to Decompose a Hundred

2 hundreds 1 ten 4 ones   1 hundred 11 tens 4 ones

Barbara is making bead necklaces. She has 253 red beads and has 314 green beads. How many beads does Barbara have?

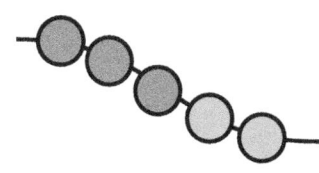

Unit Vocabulary

Adding and Subtracting within 1,000

Use this space to visualize the math vocabulary for this unit.

Word or Phrase	Example or Attributes	Visual Reminder

Unit Models & Strategies

Adding and Subtracting within 1,000

Use this space to visualize the math models and strategies for this unit.

Model or Strategy	This is a...	It is used to...

Affirmation: I don't just memorize math facts! I understand math facts!

Name: _____   Date: _____

Unit 7 Practice 1: I can use number lines to add, subtract and compare numbers.

1) Count on by tens.

540, _____, _____, _____, _____, _____, _____, _____, 620

283, _____, _____, _____, _____, _____, _____, _____, 363

920, _____, _____, _____, _____, _____, _____, _____, 1,000

2.NBT.2

2) Count back by ones.

77, _____, _____, _____, _____, _____, _____, _____, _____, 68

124, _____, _____, _____, _____, _____, _____, _____, _____, 115

556, _____, _____, _____, _____, _____, _____, _____, _____, 547

2.NBT.2

3) Solve

234 + 5 = _____   479 - 8 = _____

234 + 6 = _____   479 - 9 = _____

234 + 7 = _____   479 - 10 = _____

2.NBT.7

Do you understand?   ?   ✓

Affirmation: I don't just memorize math facts! I understand math facts!

Name: _____ Date: _____

Unit 7 Practice 1: I can use number lines to add, subtract and compare numbers.

4) Use the number line to add: 364 + 32

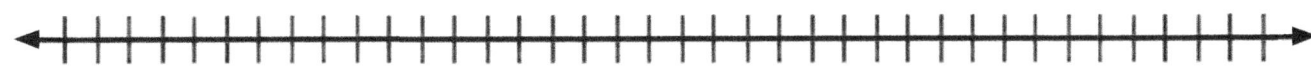

Equation: _____

2.NBT.7

5) Use the number line to subtract: 855 - 24

Equation: _____

2.NBT.7

6) Use the number line to compare 198 and 186 using >, <, or =.

198 ◯ 186

2.NBT.4

Do you understand?   ?   ✓

I Can Do Math Practice Problems, Grade 2

Affirmation: I learn from the ideas of other people, even if I don't agree with them.

Name: _____   Date: _____

Unit 7 Practice 2: I can use base-ten blocks to model three-digit numbers.

1) Count on by hundreds.

   231, _____, _____, _____, _____, _____, _____, _____, 1,031

   452, _____, _____, _____, _____, _____, _____, _____, 1,252

   404, _____, _____, _____, _____, _____, _____, _____, 1,204

   2.NBT.2

2) Count back by tens.

   720, _____, _____, _____, _____, _____, _____, _____, 640

   145, _____, _____, _____, _____, _____, _____, _____, 55

   1,000, _____, _____, _____, _____, _____, _____, _____, 920

   2.NBT.2

3) What three-digit number is modeled with these base-ten blocks?

   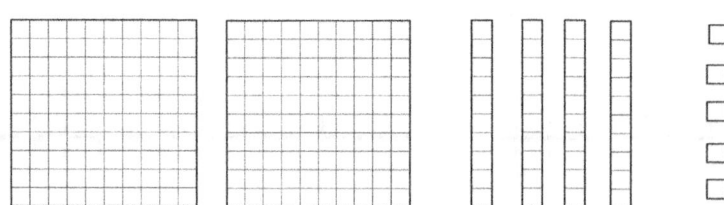

   Three-digit number: _____

   2.NBT.1

Do you understand?   ?

Name: _____ Date: _____

Unit 7 Practice 2: I can use base-ten blocks to model three-digit numbers.

4) Draw base-ten blocks to show 743.

2.NBT.1

5) Use the base-ten representation to solve 427 + 30.

Equation: _____

2.NBT.7

6) Use the base-ten representation to solve 454 - 200.

Equation: _____

2.NBT.7

Do you understand?   ?   ✓

Affirmation: I encourage myself and others by saying kind words.

Name: _____   Date: _____

Unit 7 Practice 3: I can use models to add and subtract.

1) Circle true or false for each equation.

    820 + 280 = 1,000   630 + 270 = 900   525 + 475 = 1,000

    875 - 125 = 1,000   1,000 - 150 = 850   430 + 430 = 860

2.NBT.7

2) Draw base-ten blocks to show 382.

Write 382 in **expanded form**. 382 = _____

2.NBT.3

3) Make a number line to solve: 724 + 62

Equation: _____

2.NBT.7

Do you understand? ? ✓

Affirmation: I encourage myself and others by saying kind words.

Name: _____ Date: _____

Unit 7 Practice 3: I can use models to add and subtract.

4) Make a number line to solve: 850 + _____ = 882

⟵──────────────────────────────────⟶

Equation: _____

Did you count on or count back to solve? Explain your thinking.

2.NBT.7

5) Draw base-ten blocks to solve: 245 + _____ = 365

Equation: _____

Did you count on or count back to solve? Explain your thinking.

2.NBT.7

Do you understand?   ?   ✓

Affirmation: I can learn something new every day.

Name: _____ Date: _____

Unit 7 Practice 4: I can use expanded form to add or subtract.

1) Find the missing number.

49 + _____ = 50   780 + _____ = 800

37 + _____ = 40   130 + _____ = 200

86 + _____ = 90   520 + _____ = 600

2.NBT.7

2) Find the sum.

800 + 50 + 6 = _____   100 + 90 + 1 = _____

500 + 40 + 9 = _____   700 + 2 + 30 = _____

8 + 500 + 60 = _____   200 + 5 = _____

2.NBT.3

3) Write each three-digit number in **expanded form**.

542 = _____   371 = _____

176 = _____   444 = _____

402 = _____   680 = _____

2.NBT.3

Do you understand?   ?   ✓

Affirmation: I can learn something new every day.

Name: _____ Date: _____

Unit 7 Practice 4: I can use expanded form to add or subtract.

4) Use expanded form to solve: 623 + 375

Equation: _____

2.NBT.8

5) **Agree or Disagree**:
Serenity and Arthur were solving 431 + 146 using expanded form. They wrote:

Serenity's Way	Arthur's Way
400 + 100	1 + 6
30 + 40	30 + 40
1 + 6	400 + 100

Arthur thinks they will get the same sum. Do you agree or disagree with Arthur?

I _____ with Arthur.

2.NBT.8

Do you understand? ? ✓

Affirmation: By helping others, I help myself learn more.

Name: _____ Date: _____

Unit 7 Practice 5: I can use place value to find the next hundred.

1) Match to find the number that makes each equation true.

500 = [ ] + 430   53

276 + [ ] = 300   70

347 + [ ] = 400   24

2.NBT.7

2) Make a number line to solve: 378 + _____ = 400

Equation: _____

2.NBT.7

3) Make a number line to solve: 600 - 185 = _____

Equation: _____

2.NBT.7

Do you understand? ? ✓

Affirmation: By helping others, I help myself learn more.

Name: _____ Date: _____

Unit 7 Practice 5: I can use place value to find the next hundred.

4) Draw a **base-ten representation** to solve: 398 - 145

Equation: _____

2.NBT.7

5) Use **expanded form** to solve: 381 + 128

Equation: _____

2.NBT.7

6) Make a **number line** to solve: 823 + _____ = 1,000

Equation: _____

2.NBT.7

Do you understand?   ?   ✓

Affirmation: I like thinking about harder math because it helps me become smarter.

Name: _____   Date: _____

Unit 7 Practice 6: I can compose a ten when adding.

1) Find the sums of ten.

   10 = 8 + _____   10 = _____ + 7   9 + _____ = 10

   10 = 5 + _____   10 = _____ + 4   3 + _____ = 10

   10 = 6 + _____   10 = _____ + 2   1 + _____ = 10

   K.OA.4

2) Find the missing number to compose a ten.

   24 + _____ = 30   83 + _____ = 90   9 + _____ = 20

   15 + _____ = 20   47 + _____ = 50   3 + _____ = 60

   36 + _____ = 40   18 + _____ = 20   1 + _____ = 40

   1.NBT.4

3) Add a 1-digit number to a 3-digit number.

   173 + 9 = _____   528 + 3 = _____

   436 + _____ = 440   315 + 5 = _____

   1.OA.6

Do you understand?   ?   ✓

Affirmation: I like thinking about harder math because it helps me become smarter.

Name: _____   Date: _____

Unit 7 Practice 6: I can compose a ten when adding.

4) Use the base-ten blocks to compose a ten.

   Write the 3-digit number: _____

   2.NBT.5

5) Draw base-ten blocks to show 3 hundreds, 2 tens, 18 ones.

   Write the 3-digit number: _____

   2.NBT.5

6) Draw base-ten blocks to show 1 hundred, 0 tens, 11 ones.

   Write the 3-digit number: _____

   Did you compose a ten? Yes or No

   2.NBT.5

Do you understand?   ?   ✓

Affirmation: Today is another opportunity to help someone else learn.

Name: _____   Date: _____

Unit 7 Practice 7: I can compose a ten or compose a hundred when adding.

1) Sums of 100.

        100 = 40 + _____   94 + _____ = 100

        100 = 80 + _____   63 + _____ = 100

        100 = 30 + _____   55 + _____ = 100

1.NBT.4

2) Add to compose a ten.

        36 + 14 = _____   12 + 18 = _____

        36 + 15 = _____   5 + 25 = _____

        36 + 16 = _____   23 + 27 = _____

1.NBT.4

3) Add to compose a hundred.

        73 + 31 = _____   173 + 31 = _____

        46 + 62 = _____   346 + 62 = _____

        15 + 94 = _____   215 + 94 = _____

2 NBT.5

Do you understand?   ?   ✓

I Can Do Math Practice Problems, Grade 2

Affirmation: Today is another opportunity to help someone else learn.

Name: _____ Date: _____

Unit 7 Practice 7: I can compose a ten or compose a hundred when adding.

4) Solve the expression 134 + 73.

   Did you need to compose a ten? **Yes** or **No**

   Did you need to compose a hundred? **Yes** or **No**

   2.NBT.7

5) **Agree or Disagree.**
   Sebastian and Maya are solving the expression 472 + 18.
   Sebastian thinks they will need to compose a ten.
   Maya thinks they will need to compose a hundred.

   Do you agree with Sebastian or with Maya?

   I agree with _____.

   2.NBT.7

Do you understand?   ?   ✓

Affirmation: What matters is that I tried, even if I was the last one to finish.

Name: _____ Date: _____

Unit 7 Practice 8: I can compose a ten and a hundred to add 3-digit numbers.

1) Draw base-ten blocks to show 4 hundreds, 4 tens, 13 ones.

   Use your base-ten blocks to show how you can make a ten.

   The 3-digit number is _____.

   2.NBT.3

2) Draw base-ten blocks to show 2 hundreds, 13 tens, 7 ones.

   Use your base-ten blocks to show how you can make a hundred.

   The 3-digit number is _____.

   2.NBT.3

3) Write 478 in **expanded form**.

   478 = _____

   Explain your thinking.

   2.NBT.3

Do you understand?   ?   ✓

Affirmation: What matters is that I tried, even if I was the last one to finish.

Name: _____ Date: _____

Unit 7 Practice 8: I can compose a ten and a hundred to add 3-digit numbers.

4) Find the sum.

138 + 62

Equation: _____

2.NBT.7

5) Find the sum.

574 + 34

Equation: _____

2.NBT.7

6) Find the sum.

267 + 55

Equation: _____

2.NBT.7

Do you understand?   ?   ✓

Affirmation: I should share my thoughts because it might help someone else.

Name: _____ Date: _____

Unit 7 Practice 9: I can add two 3-digit numbers.

1) Show 423 using base-ten blocks.

   Write 423 in **expanded form**.

   423 = _____

   2.NBT.3

2) Show 3 hundreds, 6 tens, 12 ones using base-ten blocks.

   Write the 3-digit number your base-ten blocks represent.

   The 3-digit number is: _____
   Explain your thinking.

   2.NBT.3

3) Match to find the sum.

   274 + 6 = [   ]   [212]

   273 + 6 = [   ]   [280]

   192 + 20 = [   ]   [279]

   2.NBT.5

Do you understand?   ?   ✓

Affirmation: I should share my thoughts because it might help someone else.

Name: _____   Date: _____

Unit 7 Practice 9: I can add two 3-digit numbers.

4) Find the sum.

$$189 + 176$$

Equation: _____

2.NBT.7

5) Dylan used expanded form to find the sum of 325 + 136. His work is shown below.

$$325 = 300 + 20 + 5$$
$$136 = 100 + 30 + 6$$
$$300 + 100 = 400$$
$$20 + 30 = 50$$
$$5 + 6 = 11$$
$$400 + 50 + 1 = 451$$

Find the error in Dylan's work. Write the correct sum.

2.NBT.9

Do you understand?   ?   ✓

Affirmation: I can do harder math problems by using math manipulatives.

Name: _____   Date: _____

Unit 7 Practice 10: I can find sums to 1,000 using different strategies.

1) Count on by hundreds.

   358, _____, _____, _____, _____, _____, _____, _____, 1,158

   101, _____, _____, _____, _____, _____, _____, _____, _____

   515, _____, _____, _____, _____, _____, _____, _____, _____

   2.NBT.2

2) Circle true 👍 or false 👎 for each equation.

   204 + 106 = 310   712 + 138 = 840   186 + 285 = 471

   340 + 106 = 350   325 + 325 = 650   195 + 305 = 400
       👍

   2.NBT.5

3) Use the base-ten representation to show the 3-digit number in different ways.

   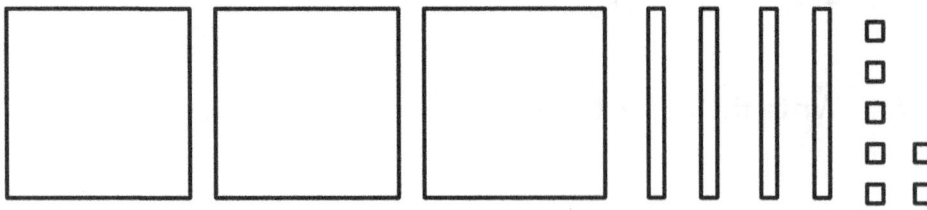

   Expanded form: _____

   Standard form: _____

   2.NBT.3

Do you understand?

Affirmation: I can do harder math problems by using math manipulatives.

Name: _____ Date: _____

Unit 7 Practice 10: I can find sums to 1,000 using different strategies.

4) What 3-digit number does the base-ten drawing show?

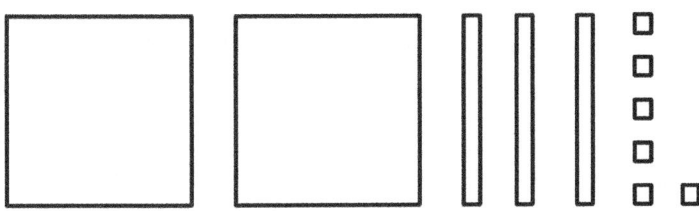

The 3-digit number is: _____

Add 476 to the base-ten drawing.

Equation: _____

2.NBT.7

5) Solve using two *different* strategies. Name the strategy you used.

639 + 249	639 + 249
Strategy: _____	Strategy: _____

2.NBT.7

Do you understand?   ?   ✓

Affirmation: I am a proficient problem solver!

Name: _____ Date: _____

Unit 7 Practice 11: I can find the sum of two 3-digit numbers.

1) Add to find the 3-digit number.

   400 + 60 + 2 = _____   20 + 80 + 120 = _____

   100 + 80 + 15 = _____   300 + 100 + 40 + 23 = _____

   500 + 40 + 200 = _____   35 + 35 + 70 + 200 = _____

   1.NBT.4

2) Find the missing number.

   500 + _____ = 900   1,000 = _____ + 600

   620 + _____ = 700   800 = _____ + 300

   280 + _____ = 300   500 = _____ + 430

   2.NBT.5

3) Write the 3-digit number **eight hundred, sixty-seven** in different ways.

   Base-ten numerals: _____

   Expanded form: _____

   Standard form: _____

   2.NBT.3

Do you understand? ? ✓

Affirmation: I am a proficient problem solver!

Name: _____ Date: _____

Unit 7 Practice 11: I can find the sum of two 3-digit numbers.

4) Use the number line to find the sum: 324 + 126

Equation: _____

2.NBT.7

5) Find the sum: 287 + 326

Equation: _____

2.NBT.7

6) Find the sum: 471 + 199

Equation: _____

2.NBT.7

Do you understand?   ?   ✓

Affirmation: My ideas are worthy of sharing.

Name: _____ Date: _____

Unit 7 Practice 12: I can decompose a ten to subtract.

1) Show 465 using base-ten blocks.

2.NBT.1

2) Subtract to find the difference.

24 - 4 = _____   78 - 9 = _____   147 - 7 = _____

24 - 5 = _____   48 - 9 = _____   147 - 8 = _____

24 - 6 = _____   68 - 9 = _____   247 - 8 = _____

2.NBT.5

3) Use the base-ten blocks to solve 237 - 16.

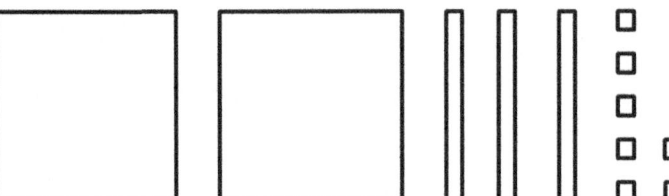

Equation: _____

2.NBT.5

Do you understand?   ?   ✓

Affirmation: My ideas are worthy of sharing.

Name: _____   Date: _____

Unit 7 Practice 12: I can decompose a ten to subtract.

4) Sample: **Decompose** a ten to solve 143 - 18.

Equation: 143 - 18 = 125

2.NBT.7

5) Use base-ten blocks to **decompose** a ten to subtract 324 - 16.

Equation: _____

Explain your thinking.

2.NBT.7

Do you understand?   ?   ✓

Affirmation: Everything learned was something we did not know at some point.

Name: _____ Date: _____

Unit 7 Practice 13: I can decompose a hundred to subtract.

1) Show 2 hundreds, 14 tens, 3 ones using base-ten blocks.

2.NBT.3

2) Subtract to find the difference.

    283 - 70 = _____   342 - 30 = _____   861 - 60 = _____

    283 - 80 = _____   342 - 40 = _____   861 - 70 = _____

    283 - 90 = _____   342 - 50 = _____   861 - 80 = _____

2.NBT.8

3) Circle true 👍 or false 👎. Decompose a hundred to subtract.

    183 - 42   431 - 12   526 - 36

    212 - 22   563 - 51   818 - 20

1.NBT.4

Do you understand?   ?   ✓

Affirmation: Everything learned was something we did not know at some point.

Name: _____ Date: _____

Unit 7 Practice 13: I can decompose a hundred to subtract.

4) **Sample: Decompose** a hundred to subtract 234 - 54.

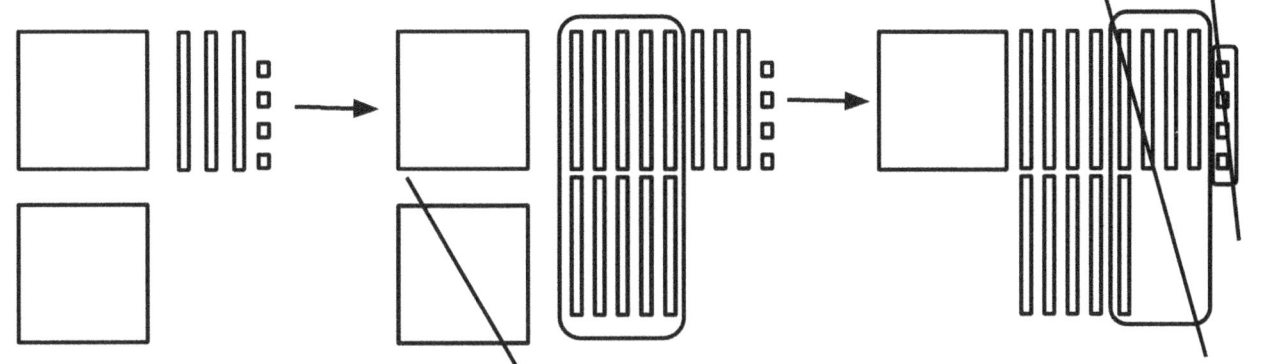

Equation: 234 - 54 = 180

2.NBT.7

5) Use base-ten blocks to **decompose** a hundred to subtract 315 - 55.

Equation:
_____

Explain your thinking.

2.NBT.7

Do you understand? ? ✓

Affirmation: Every day is a new opportunity to learn something new in math!

Name: _____ Date: _____

Unit 7 Practice 14: I can decompose a ten or a hundred to subtract.

1) Model three hundred eighty-four using base-ten blocks and in **standard form**.

   base-ten blocks

   standard form: _____

   2.NBT.3

2) Write each 3-digit number in **expanded form**.

   423 = _____   189 = _____

   691 = _____   376 = _____

   234 = _____   502 = _____

   2.NBT.3

3) Write each 3-digit number using **base-ten numerals**.

   423 equals 4 hundreds, 2 tens, 3 ones   189 equals _____

   691 equals _____   376 equals _____

   234 equals _____   502 equals _____

   2.NBT.3

Do you understand?   ?   ✓

Affirmation: Every day is a new opportunity to learn something new in math!

Name: _____ Date: _____

Unit 7 Practice 14: I can decompose a ten or a hundred to subtract.

4) **Decompose** a ten *or* a hundred to find the difference.

$$567 - 18$$

Equation: _____

I decomposed a _____ to subtract.

2.NBT.7

5) **Decompose** a ten *or* a hundred to find the difference.

$$218 - 47$$

Equation: _____

I decomposed a _____ to subtract.

2.NBT.7

Do you understand?   ?   ✓

Affirmation: Forgetting and making mistakes are a part of learning and remembering.

Name: _____   Date: _____

Unit 7 Practice 15: I can decompose a ten and a hundred to subtract.

1) Count back by tens.

   830, _____, _____, _____, _____, _____, _____, _____, _____.

   135, _____, _____, _____, _____, _____, _____, _____, _____.

   319, _____, _____, _____, _____, _____, _____, _____, _____.

   2.NBT.2

2) Sort the expressions into categories showing if you will decompose a ten *or* decompose a hundred.

   | 284 - 135 | 715 - 224 | 146 - 55 | 427 - 318 |

Decompose a ten	Decompose a hundred

   2.NBT.7

3) Find the difference: 427 - 318

   Equation: _____

   2.NBT.7

Do you understand?   ?   ✓

Affirmation: Forgetting and making mistakes are a part of learning and remembering.

Name: _____ Date: _____

Unit 7 Practice 15: I can decompose a ten and a hundred to subtract.

4) **Decompose** a ten *and* a hundred to subtract: 643 - 154

Equation: _____

2.NBT.7

5) **Decompose** a ten *and* a hundred to subtract: 158 - 99

Equation: _____

2.NBT.7

Do you understand?   ?   ✓

Affirmation: When I reason about math I get smarter.

Name: _____ Date: _____

Unit 7 Practice 16: I can use base-ten numerals to subtract 3-digit numbers.

1) Write the 3-digit numbers using base-ten numerals.

   342 equals _____   963 equals _____

   109 equals _____   814 equals _____

   780 equals _____   228 equals _____

   2.NBT.3

2) Write the missing number to make each statement true.

   3 hundreds + 4 tens + 2 ones = 2 hundreds + _____ tens + 2 ones

   1 hundred + 0 tens + 9 ones = 0 hundreds + 9 tens + _____ ones

   7 hundreds + 8 tens + 0 ones = 7 hundreds + 7 tens + _____ ones

   2.NBT.3

3) Use the base-ten blocks to subtract 300 - 87.

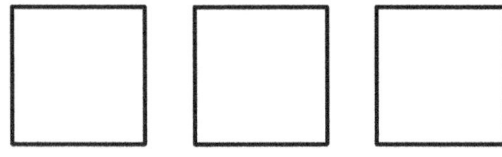

   Equation: _____

   2.NBT.7

Do you understand?   ?   ✓

Affirmation: When I reason about math I get smarter.

Name: _____ Date: _____

Unit 7 Practice 16: I can use base-ten numerals to subtract 3-digit numbers.

4) Geo wrote these steps to solve 341 - 132 using base-ten numerals.

```
                                                      3        11
Step 1:   3 hundreds    4 tens    1 one    Step 2:  3 hundreds   4̶ tens    1̶ o̶n̶e̶
      -   1 hundred     3 tens    2 ones         -   1 hundred    3 tens    2 ones
                                                    ─────────────────────────────
                                                     2 hundreds   0 tens    9 ones
```

Equation: 341 - 132 = 209

Explain the steps Geo used to find the difference.

2.NBT.9

5) Use Geo's steps to solve 341 - 152. (*Note: you may need an extra step to decompose a ten and a hundred*).

Equation: _____

2.NBT.7

Do you understand?   ?   ✓

Affirmation: I am math. From the cells in my body to the hairs on my head. I am math.

Name: _____   Date: _____

Unit 7 Practice 17: I can use a number line to subtract.

1) Subtract to find the difference.

   247 - 30 = _____   512 - 10 = _____   123 - 20 = _____

   247 - 40 = _____   512 - 20 = _____   123 - 30 = _____

   247 - 50 = _____   512 - 30 = _____   123 - 40 = _____

   2.NBT.8

2) Write the subtraction equation shown on the number line.

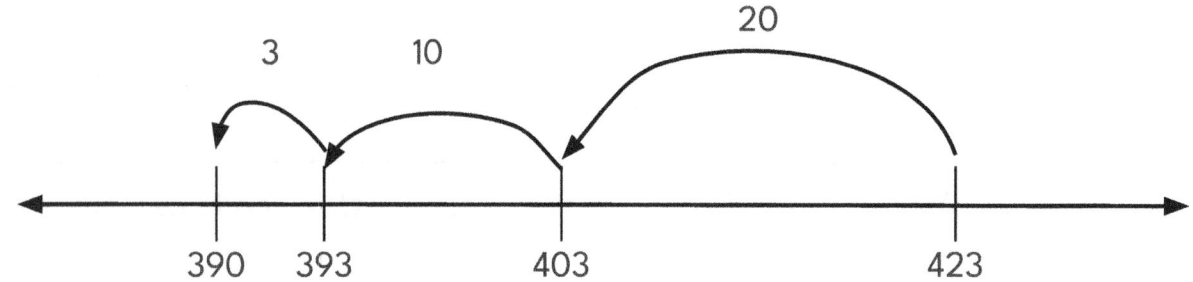

   Equation: _____

   2.NBT.8

3) Use the number line to show 314 - 41.

   Equation: _____

   2.NBT.8

Do you understand?   ?   ✓

Affirmation: I am math. From the cells in my body to the hairs on my head. I am math.

Name: _____ Date: _____

Unit 7 Practice 17: I can use a number line to subtract.

4) Use the number line to subtract: 600 - 452

Equation: _____

Explain your thinking.

2.NBT.8

5) **Agree or Disagree.** Pauline used the number line to solve the expression: 423 - 156

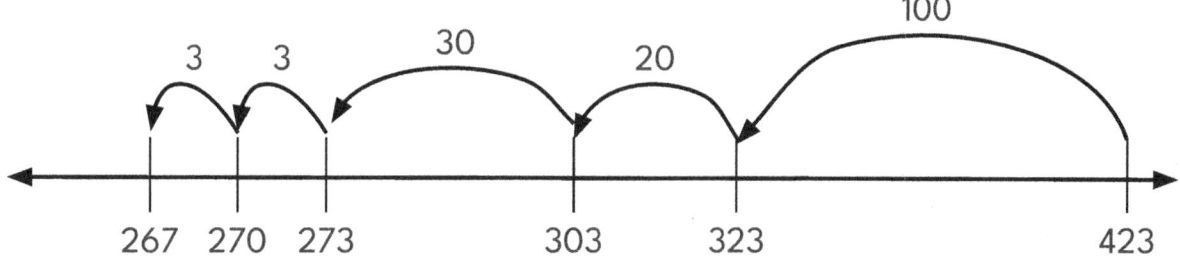

Pauline said the difference is 267. Frederick said that the difference is 156 since that is the number of jumps Pauline made on the number line.

Do you agree with Pauline or with Fredrick? I agree with _____.

2.NBT.9

Do you understand?   ?

Affirmation: When I learn something new, I will forget some of it, and that is okay.

Name: _____   Date: _____

Unit 7 Practice 18: I can use different strategies to subtract.

1) Match to find the number that makes each equation true.

200 - [ ] = 97   560

400 - [ ] = 250   103

900 - [ ] = 340   24

2.NBT.7

2) Use base-ten blocks to model 2 hundreds, 18 tens, 14 ones.

Write in **expanded form**: _____

Write the 3-digit number: _____

2.NBT.3

3) Use the number line to subtract 800 - 125.

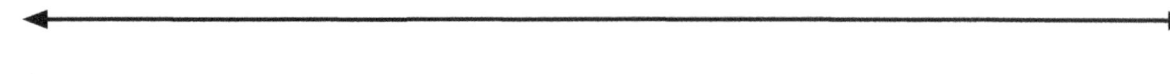

Equation: _____

2.NBT.8

Do you understand?   ?

Affirmation: When I learn something new, I will forget some of it, and that is okay.

Name: _____ Date: _____

Unit 7 Practice 18: I can use different strategies to subtract.

4) Solve using two different strategies. Name the strategy you used.

327 - 128	327 - 128

Strategy: _____   Strategy: _____

2.NBT.7

5) **Find the error.**
Ruby used base-ten numerals to solve 402 - 183. Her work is shown below.

```
Step 1:   4 hundreds   0 tens   2 one         Step 2:        9
        -  1 hundred   8 tens   3 ones                      10    12
                                                        4 hundreds   0 tens   2 ones
                                                     -  1 hundred   8 tens   3 ones
                                                        3 hundreds   1 ten    9 ones
```

Equation: 402 - 183 = 319

Find and fix the error in Ruby's work. Write the correct equation.

2.NBT.9

Do you understand? ? ✓

Affirmation: I get better the more I practice.

Name: _____ Date: _____

I can practice grade level fluencies.

Set 1: Make a ten to add

1) 25 + 5 = _____

2) 56 + 14 = _____

3) 73 + 7 = _____

4) 73 + 17 = _____

5) 45 + 6 = _____

6) 74 + 16 = _____

7) 74 + 27 = _____

2.OA.2

Set 2: Decompose a ten to subtract

1) 45 - 6 = _____

2) 31 - 4 = _____

3) 97 - 8 = _____

4) 51 - 7 = _____

5) 90 - 3 = _____

6) 40 - 5 = _____

7) 20 - 2 = _____

2.NBT.7

Do you understand?   ?   ✓

Unit Reflection

Adding and Subtracting within 1,000

Use this space to reflect on your understanding of the unit skills and concepts.

Skill/Concept	I can...	I need to work on...

Unit 8
Equal Groups

We will explore groups of objects to get ready for learning multiplication. We'll look at whether numbers are odd or even, and we'll count objects in equal rows and columns to find the total.

An Array

Equal Groups

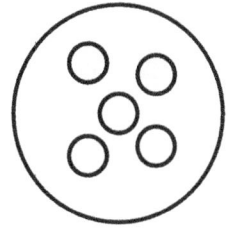
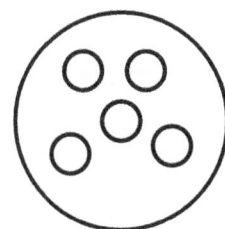

The array has 4 rows
There are 5 counters in each row

There are 2 groups
Each group has 5 counters

Even or Odd?

10 is an Even Number

Repeated Addition

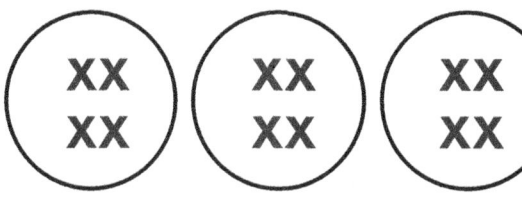

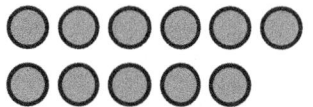

4 + 4 + 4 = 12

11 is an Odd Number

Unit Vocabulary

Equal Groups

Use this space to visualize the math vocabulary for this unit.

Word or Phrase	Example or Attributes	Visual Reminder

Unit Models & Strategies

Equal Groups

Use this space to visualize the math models and strategies for this unit.

Model or Strategy	This is a...	It is used to...

Affirmation: I am a mathematical thinker.

Name: _____ Date: _____

Unit 8 Practice 1: I can use objects to make equal groups.

1) Solve the doubles fact equation.

4 + 4 = _____   10 + 10 = _____   3 + 3 = _____

1 + 1 = _____   2 + 2 = _____   5 + 5 = _____

6 + 6 = _____   7 + 7 = _____   8 + 8 = _____

2.OA.3

2) How many counters in each group?

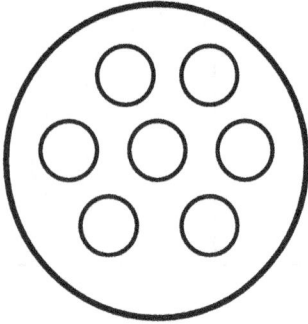   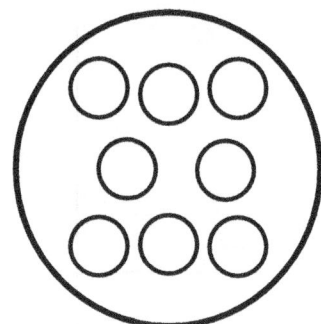

_____ counters   _____ counters

K.CC.3

3) Circle true  or false  for each equation.

10 = 5 + 5   8 = 8 + 8   18 = 9 + 9 + 1
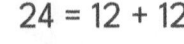   

11 = 5 + 6   24 = 12 + 12   11 = 5 + 5 + 1

2.OA.2

Do you understand? ? ✓

Affirmation: I am a mathematical thinker.

Name: _____ Date: _____

Unit 8 Practice 1: I can use objects to make equal groups.

4) Show the counters in two equal groups.

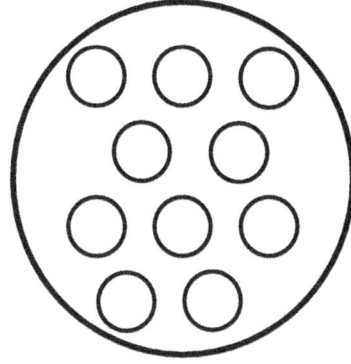

Are there any counters leftover? _____

2.OA.3

5) Archer has a collection of 17 action figures. He wants to store his collection in 2 boxes with an equal number of action figures in each box. Show how Archer can store his 17 action figures in the 2 boxes. Will he have any action figures leftover?

2.OA.3

Do you understand?   ?   ✓

Affirmation: I am brave and I take risks in math.

Name: _____ Date: _____

Unit 8 Practice 2: I can make pairs.

1) Does each counter have a pair?

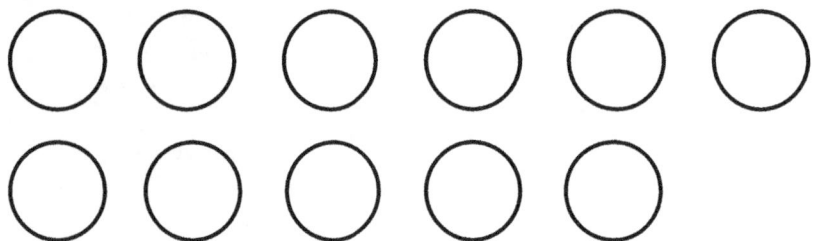

Each counter <u>does</u> / <u>does not</u> have a pair.
  circle one

K.CC.6

2) Circle the pictures that show a pair.

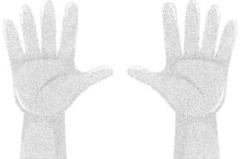

2.OA.3

3) The toy robot has a pair of eyes and a pair of legs. What other pairs can you find on the toy robot?

2.OA.3

Do you understand?   ?   ✓

Affirmation: I am brave and I take risks in math.

Name: _____ Date: _____

Unit 8 Practice 2: I can make pairs.

4) **Even or Odd?**
   Is this an even number of candies or an odd number of candies?

2.OA.3

5) Gabby has 14 counting cubes. Show how Gabby can model her counting cubes in pairs.

Is 14 an even or odd number? 14 is an <u>even / odd</u> number.
circle one

2.OA.3

Do you understand?   ?   ✓

I Can Do Math Practice Problems, Grade 2

Affirmation: Math is more than just getting an answer.

Name: ______________________ Date: __________

Unit 8 Practice 3: I can identify even and odd groups of objects.

1) Count on by 2s.

2, ______, ______, ______, ______, ______, ______, ______, ______, 20

22, ______, ______, ______, ______, ______, ______, ______, ______, 40

1,3, ______, ______, ______, ______, ______, ______, ______, ______, 21

2.NBT.2

2) Circle the dot cards that show an **odd** number of dots.

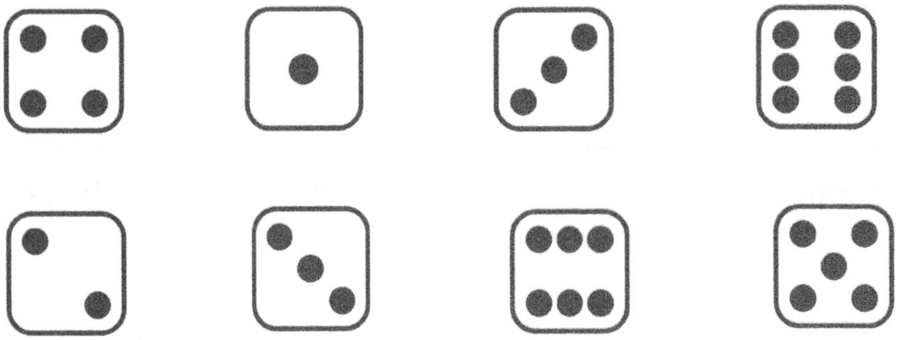

2.OA.3

3) Draw a model to show that 10 is an even or an odd number.

10 is an <u>even / odd</u> number.
circle one

2.OA.3

Do you understand? **?** ✓

Affirmation: Math is more than just getting an answer.

Name: _____ Date: _____

Unit 8 Practice 3: I can identify even and odd groups of objects.

4) The dots on a domino are called *pips*. Is there an even number or an odd number of pips on this domino?

There is an <u>even / odd</u> number of pips on this domino.
       circle one

2.OA.3

5) Xavier wants to draw a domino with an odd number of pips. The left side of his domino is shown here.

Draw pips on the right side of the domino to show two different ways Xavier could draw a domino showing an odd number of pips.

There are _____ pips on this domino.

There are _____ pips on this domino.

2.OA.3

Do you understand?

Affirmation: When I do math, I feel like a superhero.

Name: _____ Date: _____

Unit 8 Practice 4: I can decompose a number to show even or odd.

1) Match each sum with its equal number of addends.

Sum	Addends
8	6 + 6
6	4 + 4
12	3 + 3
2	2 + 2
4	1 + 1

1.OA.6

2) Draw pips on a domino to model each equation.

8 = 4 + 4

6 = 3 + 3

9 = 4 + 5

11 = 5 + 6

1.OA.6

Do you understand?   ?   ✓

Affirmation: When I do math, I feel like a superhero.

Name: _____ Date: _____

Unit 8 Practice 4: I can decompose a number to show even or odd.

3) Luke has 16 candies to put into two bags. Show how Luke can share the 16 candies in two bags with the same number of candies in each bag.

Equation: _____

Is 16 an even number or an odd number?

16 is an <u>even / odd</u> number.
        circle one

2.OA.3

4) **Agree or Disagree:**
Essence has 17 stickers. She told Donald that she can use the 17 stickers to put the same number of stickers on 2 pages in her sticker book. Donald said that she cannot put the same number of stickers on two pages in her book because 17 is an odd number.

Do you agree with Essence or with Donald?

I agree with _____.

2.OA.3

Do you understand?   ?   ✓

Affirmation: It is important that I explain and justify my thinking.

Name: _____   Date: _____

Unit 8 Practice 5: I can find patterns using even numbers and odd numbers.

1) Find the sum.

    2 + 2 = _____   5 + 5 = _____   6 + 1 = _____

    4 + 1 = _____   10 + 1 = _____   12 + 1 = _____

    3 + 3 = _____   6 + 6 = _____   8 + 7 = _____

2.OA.2

2) Sort the expressions into categories showing even or odd sums.

    3 + 3   3 + 2   2 + 3   5 + 5   6 + 4   6 + 6

Even Sums	Odd Sums

2.OA.3

3) Circle the even numbers on the number line.

What pattern do you see?

2.NBT.2

Do you understand? ? ✓

Affirmation: It is important that I explain and justify my thinking.

Name: _____ Date: _____

Unit 8 Practice 5: I can find patterns using even numbers and odd numbers.

4) Joel sorted dot cards to show even and odd numbers.

Even Numbers	Odd Numbers

What pattern do you see in Joel's dot card sort?

2.OA.3

5) **Agree or Disagree.**
Eli used 6 as his **start number**. He said that when he adds 1 to his start number, the result is an odd number but when he adds 2 to his start number, the result is an even number.

Do you agree or disagree with Eli?

I <u>agree / disagree</u> with Eli.
    circle one

2.OA.3

Do you understand? ? ✓

Affirmation: We are all smart in different ways.

Name: _____ Date: _____

Unit 8 Practice 6: I can show sums with more than two addends.

1) Count on by tens.

10, _____, _____, _____, _____, _____, _____, _____, _____, 100

Count on by 2s.

2, _____, _____, _____, _____, _____, _____, _____, _____, 20

2.NBT.2

2) Find the sums.

10 + 10 = _____   2 + 2 = _____

10 + 10 + 10 = _____   2 + 2 + 2 = _____

10 + 10 + 10 + 10 = _____   2 + 2 + 2 + 2 = _____

10 + 10 + 10 + 10 + 10 = _____   2 + 2 + 2 + 2 + 2 = _____

2.NBT.2

3) Add to find one more or two more.

12 + 2 = _____   2 + 2 = _____   8 + 1 = _____

3 + 2 = _____   5 + 1 = _____   9 + 2 = _____

16 + 1 = _____   13 + 2 = _____   4 + 2 = _____

2.OA.2

Do you understand?   ?   ✓

Affirmation: We are all smart in different ways.

Name: _____ Date: _____

Unit 8 Practice 6: I can show sums with more than two addends.

4) **Agree or Disagree.**
Carter has 2 bags of marble with 6 marbles in each bag. Heather has 6 bags of marbles with 2 marbles in each bag. Carter said that they have the same number of marbles.

Do you agree or disagree with Carter?

I <u>agree / disagree</u> with Carter.
 circle one

2.OA.4

5) Draw a model that shows 3 bags of cookies with 10 cookies in each bag.

How many cookies are in the 3 bags?

There are _____ in the three bags.

Write an equation to match your drawing.

Equation: _____

2.OA.4

Do you understand?   ?   ✓

Affirmation: Math helps me make sense of my world.

Name: _____ Date: _____

Unit 8 Practice 7: I can count rows in an array.

1) How many counters?

   ○○ ○○○
   ○○○ ○
   ○○

   There are _____ counters.

   Write an equation to show your thinking.

   Equation: _____

   K.CC.5a

2) Name the flat shapes.

   [rectangle]   [square]   [rectangle]

   _____   _____   _____

   K.G.2

3) Draw counters to show 12.

   Write an equation to show your thinking.

   Equation: _____

   K.CC.5b

Do you understand?   ?   ✓

Affirmation: Math helps me make sense of my world.

Name: _____   Date: _____

Unit 8 Practice 7: I can count rows in an array.

4) Use the **array** to answer the questions.

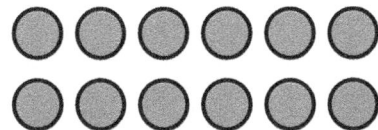

How many **rows** are there in this array? There are _____ rows in this array.

How many counters are there in each row? There are _____ counters in each row.

How many counters are in the array? There are _____ counters in the array.

2.OA.5

5) Use the **array** to answer the questions.

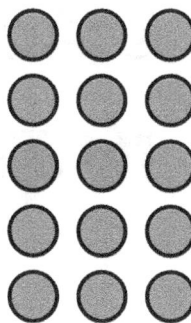

How many **rows** are there in this array? There are _____ rows in this array.

How many counters are there in each row? There are _____ counters in each row.

How many counters are in the array? There are _____ counters in the array.

2.OA.5

Do you understand?   ?   ✓

Affirmation: Acting out math problems helps me understand what I am doing.

Name: _____ Date: _____

Unit 8 Practice 8: I can count columns in an array.

1) Count on by twos.

2, _____, _____, _____, _____, _____, _____, _____, _____, 20

Count on by fives.

5, _____, _____, _____, _____, _____, _____, _____, _____, 50

2.NBT.2

2) Circle true 👍 or false 👎. The counters show an **array**.

2.OA.4

3) Draw an **array** with 15 counters.

2.OA.4

Do you understand?   ?   ✓

I Can Do Math Practice Problems, Grade 2

Affirmation: Acting out math problems helps me understand what I am doing.

Name: _____ Date: _____

Unit 8 Practice 8: I can count columns in an array.

4) Use the array to answer the questions.

How many **columns** are there in this array? There are _____ columns in this array.

How many counters are there in each column? There are _____ counters in each column.

How many counters are in the array? There are _____ counters in the array.

2.OA.4

5) Use the array to answer the questions.

How many **columns** are there in this array? There are _____ columns in this array.

How many counters are there in each column? There are _____ counters in each column.

How many counters are in the array? There are _____ counters in the array.

2.OA.4

Do you understand?   ?   ✓

Affirmation: Sketching out my thinking helps me see a problem more clearly.

Name: _____   Date: _____

Unit 8 Practice 9: I can use an expression to model an array.

1) Find the sum.

2 + 2 = _____   3 + 3 = _____

2 + 2 + 2 = _____   3 + 3 + 3 = _____

2 + 2 + 2 + 2 = _____ 3 + 3 + 3 + 3 = _____

4 + 4 = _____   5 + 5 = _____

4 + 4 + 4 = _____   5 + 5 + 5 = _____

4 + 4 + 4 + 4 = _____ 5 + 5 + 5 + 5 = _____

2.NBT.2

2) Match the array to the expression.

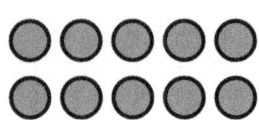      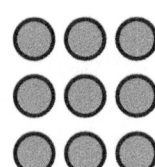   

| 3 + 3 | 5 + 5 | 2 + 2 + 2 + 2 + 2 | 3 + 3 + 3 |

2.OA.4

Do you understand?   ?   ✓

Affirmation: Sketching out my thinking helps me see a problem more clearly.

Name: _____ Date: _____

Unit 8 Practice 9: I can use an expression to model an array.

3) Write two expressions to model the array.

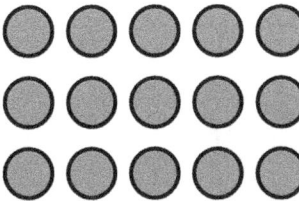

Expression: _____

Expression: _____

How many counters are in the array?

There are _____ counters in the array.

2.OA.4

4) **Agree or Disagree:**
Carly drew this array to model the expression 4 + 4 + 4.

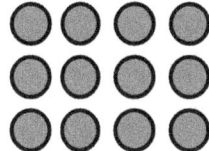

Freddy said that the expression to match Carly's array is 3 + 3 + 3 + 3.
Do you agree or disagree with Carly or with Freddy?

I agree with _____.

2.OA.4

Do you understand?   ?

Affirmation: Making mistakes is how I learn new things.

Name: _____ Date: _____

Unit 8 Practice 10: I can draw arrays and write expressions.

1) Count on by 3s.

3, _____, _____, _____, _____, _____, _____, _____, _____, 30

Count on by 4s.

4, _____, _____, _____, _____, _____, _____, _____, _____, 40

2.NBT.2

2) Find the sum.

3 + 3 = _____   4 + 4 = _____   5 + 5 = _____

6 + 3 = _____   8 + 4 = _____   10 + 5 = _____

9 + 3 = _____   12 + 4 = _____   15 + 5 = _____

2.OA.2

3) How many counters in the array?

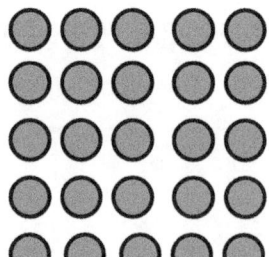

There are _____ counters in the array.

2.OA 4

Do you understand?   ?   ✓

Affirmation: Making mistakes is how I learn new things.

Name: _____ Date: _____

Unit 8 Practice 10: I can draw arrays and write expressions.

4) Draw an array with 12 counters.

My array has _____ rows.

My array has _____ columns.

Write an expression to match your array.

Expression: _____

2.OA.4

5) Draw an array to match the expression 6 + 6 + 6.

My array has _____ rows.

My array has _____ columns.

Write a different expression that also models you array.

Expression: _____

2.OA.4

Do you understand? ? ✓

Affirmation: I can help others by asking questions.

Name: _____ Date: _____

Unit 8 Practice 11: I can write an equation to model an array.

1) Match the expressions that have the same sum.

2 + 2 + 2		3 + 3 + 3 + 3 + 3
5 + 5 + 5		3 + 3
6 + 6		2 + 2 + 2 + 2
4 + 4		2 + 2 + 2 + 2 + 2 + 2

2.NBT.5

2) The rectangle is split into 4 equal pieces.

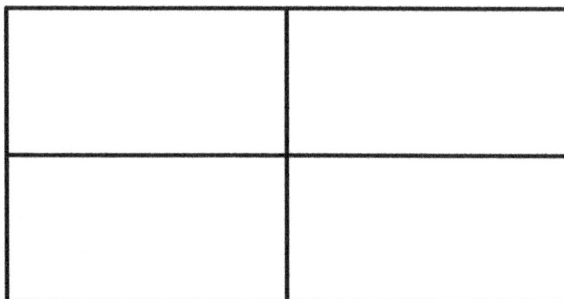

Draw two vertical lines to split the rectangle into 8 equal pieces.

What shape is each of the equal pieces inside the rectangle?

Each equal piece is a _____.

2.G.2

Do you understand?   ?   ✓

Affirmation: I can help others by asking questions.

Name: _____ Date: _____

Unit 8 Practice 11: I can write an equation to model an array.

3) Fill the empty rectangle with squares to match the array built with counters.

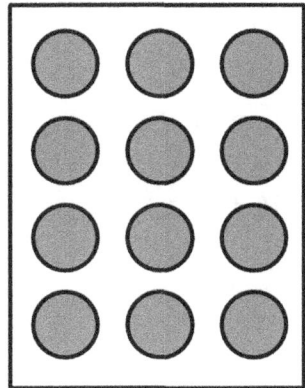 

Write an equation that models the array you made with squares.

Equation: _____

2.G.2

4) **Find the error:**
Sammy drew this array to model the equation 5 + 5 + 5 = 15.

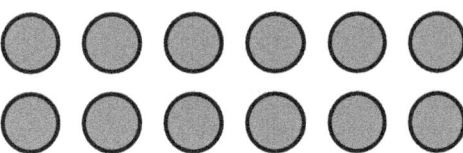

Find and fix the error in Sammy's work.

2.OA.4

Do you understand? ? ✓

Affirmation: I can do harder math problems by acting out the story.

Name: _____ Date: _____

Unit 8 Practice 12: I can fill rectangles with equal size squares to model arrays.

1) Solve using doubles facts.

      3 + 3 = _____   5 + _____ = 10   10 + 10 = _____

      5 + 5 = _____   7 + _____ = 15   12 + 12 = _____

      9 + 9 = _____   2 + _____ = 5   14 + 14 = _____

      6 + 6 = _____   4 + _____ = 7   16 + 16 = _____

2.NBT.5

2) How many squares are in the rectangle?

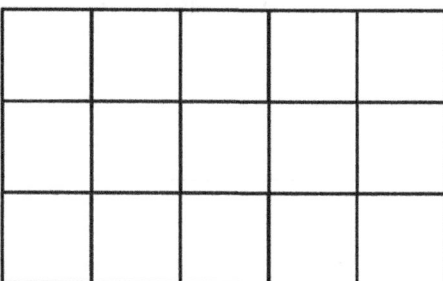

There are _____ squares inside the rectangle.

2.G.2

3) Draw squares inside the rectangle to model 2 + 2 + 2.

2.G.2

Do you understand?   ?   ✓

Affirmation: I can do harder math problems by acting out the story.

Name: _____ Date: _____

Unit 8 Practice 12: I can fill rectangles with equal size squares to model arrays.

4) Draw lines to finish filling the rectangle with squares.

There are _____ squares in the rectangle.

2.G.2

5) **Agree or Disagree:**
Dennis and Jade made this array with square tiles.

Dennis said that the square tiles show an array inside a rectangle.
Jade said that the square tiles made an array inside a square.

Do you agree with Dennis or with Jade?

I agree with _____.

2.OA.4

Do you understand? ? ✓

Affirmation: I am a problem solver.

Name: _____ Date: _____

Unit 8 Practice 13: I can find sums and differences and partition rectangles.

1) Count back by tens.

100, _____, _____, _____, _____, _____, _____, _____, _____, 10

Count back by fives.

50, _____, _____, _____, _____, _____, _____, _____, _____, 5

2.NBT.2

2) Find the sum.

$24 + 10 =$ _____ $87 + 12 =$ _____ $35 + 5 =$ _____

$37 + 20 =$ _____ $43 + 12 =$ _____ $45 + 5 =$ _____

$56 + 20 =$ _____ $65 + 5 =$ _____ $65 + 15 =$ _____

2.NBT.8

3) Find the difference.

$93 - 10 =$ _____ $93 - 12 =$ _____ $40 - 5 =$ _____

$93 - 20 =$ _____ $34 - 12 =$ _____ $65 - 15 =$ _____

$100 - 10 =$ _____ $104 - 12 =$ _____ $17 - 5 =$ _____

2.NBT.7

Do you understand?   ?   ✓

Affirmation: I am a problem solver.

Name: _____ Date: _____

Unit 8 Practice 13: I can find sums and differences and partition rectangles.

4) Partition the rectangle into equal-size squares to create an array.

Write an equation to model your array.

Equation: _____

2.G.2

5) Draw lines to finish filling the rectangle with 8 equal-size squares.

How are there arrays the same?

How are these arrays different?

2.OA.4

Do you understand?   ?   ✓

Affirmation: I get better the more I practice.

Name: _____ Date: _____

I can practice grade level fluencies.

Set 1: Find the sum

1) 6 + 6 = _____

2) 9 + 9 = _____

3) 10 + 10 = _____

4) 7 + 7 = _____

5) 11 + 11 = _____

6) 14 + 14 = _____

7) 15 + 15 = _____

2.OA.3

Set 2: Find the sum

1) 4 + 4 = _____

2) 8 + 8 = _____

3) 5 + 5 = _____

4) 12 + 12 = _____

5) 16 + 16 = _____

6) 13 + 13 = _____

7) 19 + 19 = _____

2.OA.3

Do you understand?   ?   ✓

Unit Reflection

Equal Groups

Use this space to reflect on your understanding of the unit skills and concepts.

Skill/Concept	I can...	I need to work on...

Unit 9
Putting It All Together

We will keep getting better at adding and subtracting numbers quickly and carefully. We will also solve story problems using different strategies and what we know about place value.

Addition
132 + 246 = 378

Subtraction
439 - 127 = 312

Base-Ten Numerals

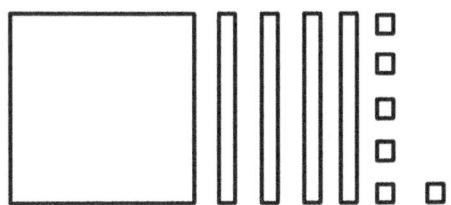

146 has 1 hundred, 4 tens, 6 ones

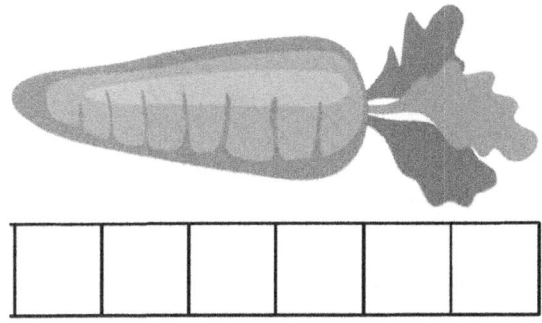

The carrot is about 6 tiles long

Story Problem:
Jo-Jo has a ribbon that is 68 inches long.
Jo-Jo cut off 36 inches of the ribbon to wrap a package.
How much ribbon does Jo-Jo have now?

68 - 36 = ☐

Jo-Jo now has 32 inches of ribbon.

Unit Vocabulary

Putting It All Together

Use this space to visualize the math vocabulary for this unit.

Word or Phrase	Example or Attributes	Visual Reminder

Unit Models & Strategies

Putting It All Together

Use this space to visualize the math models and strategies for this unit.

Model or Strategy	This is a…	It is used to…

Affirmation: If I messed up yesterday, today is a new day, and I can try again.

Name: _____ Date: _____

Unit 9 Practice 1: I can add and subtract within 20.

1) The number bond shows 10 = 9 + 1. This is one way you can break-apart 10. Use the equations to show all the other ways you can break-apart 10.

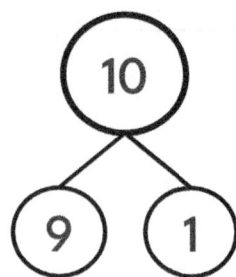

10 = _____ + _____   10 = _____ + _____

10 = _____ + _____   10 = _____ + _____

10 = _____ + _____   10 = _____ + _____

10 = _____ + _____   10 = _____ + _____

K.OA.3

2) Find the missing number.

10 - _____ = 8   11 - _____ = 9   12 - _____ = 3

10 - _____ = 3   11 - _____ = 5   12 - _____ = 4

10 - _____ = 6   11 - _____ = 8   12 - _____ = 7

10 - _____ = 5   11 - _____ = 7   12 - _____ = 6

1.OA.4

3) Use doubles facts to solve.

3 + 3 = _____   8 + 8 = _____   6 + 6 = _____

3 + 4 = _____   9 + 8 = _____   6 + 5 = _____

4 + 4 = _____   9 + 9 = _____   5 + 5 = _____

5 + 4 = _____   10 + 9 = _____   4 + 5 = _____

1.OA.6

Do you understand?   ?   ✓

Affirmation: If I messed up yesterday, today is a new day, and I can try again.

Name: _____ Date: _____

Unit 9 Practice 1: I can add and subtract within 20.

4) Make a ten. Find the missing number.

$9 + 3 = 9 + 1 +$ _____ $6 + 8 = 6 + 4 +$ _____

$8 + 4 = 8 + 2 +$ _____ $6 + 8 =$ _____ $+ 2 + 8$

$7 + 9 =$ _____ $+ 1 + 9$ $5 + 7 =$ _____ $+ 3 + 7$

1.OA.6

5) **Agree or Disagree.**
Rhyan said that knowing $10 = 4 + 6$ helped him to solve $20 - 4$. Rhyan wrote the following to show his thinking.

$20 - 4 =$
$10 - 4 = 6$
$10 + 6 = 16$
So, $20 - 4 = 16$

Do you agree or disagree with Rhyan?

I _____ with Rhyan.

Show another way you can solve $20 - 4$.

2.OA.2

Do you understand?   ?   ✓

I Can Do Math Practice Problems, Grade 2

Affirmation: Math helps me develop grit and perseverance.

Name: _____ Date: _____

Unit 9 Practice 2: I can use addition or subtraction to find sums and differences.

1) Match to find the missing addend.

17 + ☐   8

6 + ☐   2

3 + ☐   4

12 + ☐   9

1.OA.8

2) Find the difference.

20 - _____ = 14   18 - _____ = 7

20 - _____ = 8   19 - _____ = 9

20 - _____ = 16   16 - _____ = 8

20 - _____ = 11   17 - _____ = 6

20 - _____ = 3   15 - _____ = 9

1.OA.4

Do you understand?   ?   ✓

Affirmation: Math helps me develop grit and perseverance.

Name: _____ Date: _____

Unit 9 Practice 2: I can use addition or subtraction to find sums and differences.

3) Solve using two different strategies. Name the strategy you used.

8 + 4 + 7   8 + 4 + 7

Strategy: _____   Strategy: _____

2.OA.2

4) **Find the error:**
Hallie used a doubles fact to solve 9 + 8 + 5. Her thinking is below.

9 + 8 + 5 = ?
9 + 9 = 18
18 + 5 = 23
So, 9 + 8 + 5 = 23

Find and fix the error in Hallie's work.

2.OA.2

Do you understand?   ?   ✓

Affirmation: I don't need to be fast at math. I need to understand and that takes time.

Name: _____ Date: _____

Unit 9 Practice 3: I can add or subtract on a grid to find distance.

1) Circle true or false for each equation.

$5 + 7 = 12 - 0$ $20 - 7 = 7 + 13$

$8 - 3 = 8 + 3$ $20 - 8 = 4 + 8$

$9 - 4 = 4 + 5$ $11 + 5 = 10 + 6$

1.OA.3

2) Use the base-ten cubes to measure the length of each object.

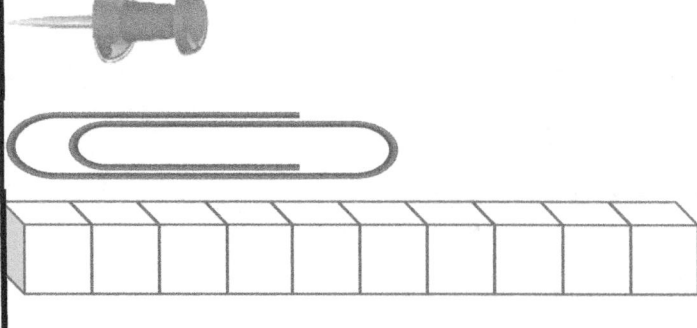

The push pin is _____ base-ten cubes long.

The paperclip is _____ base-ten cubes long.

2.MD.4

Do you understand?   ?   ✓

Affirmation: I don't need to be fast at math. I need to understand and that takes time.

Name: _____   Date: _____

Unit 9 Practice 3: I can add or subtract on a grid to find distance.

Daman's Walking Distance

• house   • library   • school

• park

A) Each box on the grid is one unit. Find the distance for each location in units.

How many units is it from Daman's house to the school?

The distance from Daman's house to the school is _____ units.

How many units is it from Daman's house to the park?

The distance from Daman's house to the park is _____ units.

2.MD.3

B) What is the difference in the distance from Daman's house to the school and the distance from Daman's house to the library?

Equation: _____

2.MD.4

Do you understand?

I Can Do Math Practice Problems, Grade 2

Affirmation: Think positive thoughts. Negative thoughts don't help us learn and grow.

Name: _____ Date: _____

Unit 9 Practice 4: I can use measurement data to make a line plot.

1) Find the sum.

    24 + 62 = _____   35 + 36 = _____   49 + 26 = _____

    67 + 23 = _____   52 + 56 = _____   84 + 15 = _____

    17 + 22 = _____   75 + 24 = _____   31 + 29 = _____

    2.NBT.5

2) Find the difference.

    64 - 31 = _____   46 - 17 = _____   76 - 25 = _____

    78 - 25 = _____   100 - 46 = _____   51 - 25 = _____

    97 - 46 = _____   23 - 12 = _____   92 - 38 = _____

    2.NBT.5

3) Find the sum.

    3 + 3 = _____   4 + 4 = _____   6 + 6 = _____

    3 + 3 + 3 = _____   4 + 4 + 4 = _____   12 + 6 = _____

    3 + 3 + 3 + 3 = _____   4 + 4 + 4 + 4 = _____   18 + 6 = _____

    2.NBT.2

Do you understand?   ?   ✓

Affirmation: Think positive thoughts. Negative thoughts don't help us learn and grow.

Name: _____   Date: _____

Unit 9 Practice 4: I can use measurement data to make a line plot.

Questions A and B refer to the information below.   Complete the line plot using data shown.

Length of Paper Clips in cm	Number of paperclips
4 cm	3
8 cm	4
6 cm	2
5 cm	3

Arnie's Paper Clip Collection

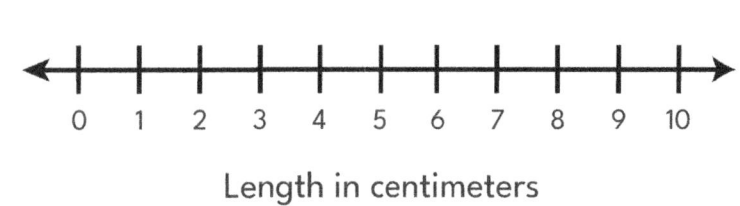

Length in centimeters

A) How many paper clips does Arnie have in his collection?

Equation: _____

Arnie has _____ paperclips in his collection.

2.MD.9

B) Arnie made a paper clip chain placing the paper clips that measure 5 centimeters end-to-end without gaps or overlaps. How long is Arnie's paper clip chain?

Equation: _____

Arnie's paperclip chain is _____ centimeters long.

2.MD.9

Do you understand?   ?

Affirmation: Mathematics makes me smarter because it makes me think!

Name: _____   Date: _____

Unit 9 Practice 5: I can show 3-digit numbers in different ways.

1) How many hundreds, tens, ones are shown here?

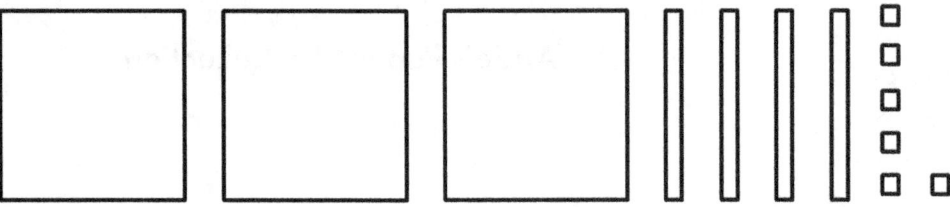

There are _____ hundreds, _____ tens, _____ ones.

The 3-digit number is _____.

2.NBT.1

2) Show the 3-digit number 523 using base-ten blocks.

2.NBT.1

3) Show the number two hundred fifty-nine using base-ten blocks.

2.NBT.3

Do you understand?   ?   ✓

Affirmation: Mathematics makes me smarter because it makes me think!

Name: _____ Date: _____

Unit 9 Practice 5: I can show 3-digit numbers in different ways.

4) Draw base-ten blocks to show the number 107 in two different ways.

          107 | 107

2.NBT.3

5) Rama used base-ten blocks to show a 3-digit number.

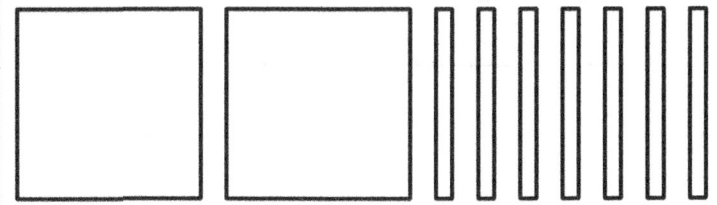

How many of each base-ten blocks did Rama use?

Rama used _____ hundreds, _____ tens, _____ ones.

What is Rama's 3-digit number?

Rama's 3-digit number is _____.

2.NBT.3

Do you understand?

Affirmation: I am human and humans are mathematical beings.

Name: _____   Date: _____

Unit 9 Practice 6: I can write equations using equal expressions.

1) Write the 3-digit number shown in expanded form to make the equation true.

   200 + 40 + 5 = _____   800 + 50 + 3 = _____

   20 + 9 + 500 = _____   400 + 80 + 2 = _____

   100 + 3 = _____   20 + 900 + 3 = _____

   2.NBT.3

2) Match the 3-digit number to its word form.

456		four hundred five
465		four hundred fifty-six
405		four hundred sixty-five

   2.NBT.3

3) Write the 3-digit number represented in base-ten numerals.

   2 hundreds, 6 tens, 9 ones is _____.   18 ones, 1 ten, 3 hundreds is _____.

   7 tens, 3 ones, 8 hundreds is _____.   45 tens, 6 ones is _____.

   5 hundreds, 14 tens, 1 one is _____.   6 hundreds, 0 tens, 17 ones is _____.

   2.NBT.3

Do you understand?   ?   ✓

Affirmation: I am human and humans are mathematical beings.

Name: _____ Date: _____

Unit 9 Practice 6: I can write equations using equal expressions.

4) Finish the expression to make the equation true.

   1 hundred + 6 tens + 15 ones = 1 hundred + _____ tens 5 ones

   5 hundreds + 18 tens + 3 ones = _____ hundreds + 8 tens + 3 ones

   2 hundreds + 19 tens + 18 ones = _____ hundreds + 0 tens + 8 ones

   4 hundreds + 7 tens + 2 ones = 3 hundreds + _____ tens + _____ ones

   2.NBT.1

5) Kamaji has only tens and ones in base-ten blocks. Draw base-ten blocks to show how Kamaji can represent 1 hundred, 4 tens, 15 ones using only tens and ones.

   Finish the expression that Kamaji's showed using base-ten blocks.

   1 hundred + 4 tens + 15 ones = _____ tens + _____ ones

   2.NBT.1

Do you understand?   ?

Affirmation: Even when we know the answer, it can be fun to try out a different way.

Name: _____ Date: _____

Unit 9 Practice 7: I can use different strategies to add and subtract within 1,000.

1) Count on by hundreds.

245, _____, _____, _____, _____, _____, _____, _____

Count on by fives.

95, _____, _____, _____, _____, _____, _____, _____

2.NBT.2

2) Count back by hundreds.

1,234, _____, _____, _____, _____, _____, _____, _____

Count back by fives.

115, _____, _____, _____, _____, _____, _____, _____

2.NBT.2

3) Circle true  or false  for each equation.

| 627 + 210 = 837 | 403 - 102 = 301 | 431 + 132 = 663 |
| 200 - 190 = 110 | 575 + 125 = 600 | 120 - 15 = 105 |

2.NBT.8

Do you understand?   ?   ✓

Affirmation: Even when we know the answer, it can be fun to try out a different way.

Name: _____   Date: _____

Unit 9 Practice 7: I can use different strategies to add and subtract within 1,000.

4) Solve using two different strategies. Name the strategy you used.

| 349 + 267 | 349 + 267 |

Strategy: _____   Strategy: _____

2.NBT.7

5) Solve using two different strategies. Name the strategy you used.

| 741 - 325 | 741 - 325 |

Strategy: _____   Strategy: _____

2.NBT.7

Do you understand?   ?   ✓

Affirmation: I have a powerful mind!

Name: _____   Date: _____

Unit 9 Practice 8: I can add and subtract within 100 using mental math.

1) Write the missing number that makes the equation true.

    5 + _____ = 14   75 + _____ = 100   32 + _____ = 40

    12 + _____ = 20   8 + _____ = 50   15 + _____ = 33

    17 + _____ = 30   13 + _____ = 25   9 + _____ = 50

1.OA.4

2) Match the subtraction equation to the number that shows the difference.

    50 - 26 = ☐   20

    82 - 62 = ☐   21

    44 - 23 = ☐   24

2.NBT.5

3) Find the sum or the difference using multiples of ten.

    57 + _____ = 67   48 - _____ = 38   19 + _____ = 29

    26 - _____ = 6   65 - _____ = 35   22 + _____ = 52

    83 + _____ = 103   105 - _____ = 55   37 + _____ = 77

2.NBT.8

Do you understand? ? ✓

Affirmation: I have a powerful mind!

Name: _____ Date: _____

Unit 9 Practice 8: I can add and subtract within 100 using mental math.

4) Lindie was solving 47 - 29.
   Lindie said that knowing 47 - 30 = 17 can help to solve the problem.

   How can knowing 47 - 30 = 17 help Lindie solve 47 - 29?

   Equation: 47 - 29 = _____

   2.NBT.8

5) **Find the error:**
   Dori used counting on by ten to solve 37 + 40. Her thinking is below.

   Count on by ten starting at 37.
   37, 47, 57, 67

   So, 37 + 40 = 67

   Find and fix the error in Dori's work.

   2.NBT.8

Do you understand?   ?   ✓

Affirmation: I am brilliant, bright, and getting better every day!

Name: _____ Date: _____

Unit 9 Practice 9: I can solve different types of story problems.

1) Find the sum of three addends.

    $3 + 6 + 7 =$ _____ $19 + 4 + 5 =$ _____

    $7 + 7 + 3 =$ _____ $12 + 4 + 4 =$ _____

    $19 = 8 + 1 +$ _____ $12 = 5 + 2 +$ _____

    1.OA.2

2) There are 12 red fish and 7 yellow fish in a fishtank. How many fish are in the fishtank?

    Equation: _____

    1.OA.1

3) There are 28 pencils in the class pencil bin. Twelve of the pencils are sharpened. How many pencils in the bin are not sharpened?

    Equation: _____

    1.OA.1

Do you understand?   ?   ✓

Affirmation: I am brilliant, bright, and getting better every day!

Name: _____ Date: _____

Unit 9 Practice 9: I can solve different types of story problems.

4) Nevya has a collection of shapes. Nevya's collection has 8 circles, 3 triangles and 6 squares. How many shapes are in Nevya's collection?

Equation: _____

1.OA.2

5) A local zoo has a reptile enclosure with 17 geckos and 9 snakes. How many more geckos than snakes are there in the zoo's reptile enclosure?

Equation: _____

2.OA.1

Do you understand?   ?   ✓

Name: _____ Date: _____

Unit 9 Practice 10: I can write a question from a story problem.

1) Write the missing symbols (+, - , and =) to make a true equation.

25 ☐ 10 ☐ 15 | 18 ☐ 12 ☐ 30

24 ☐ 26 ☐ 50 | 87 ☐ 44 ☐ 43

1.OA.3

2) Write a number that is less than 41.

_____ is less than 41.

1.NBT.3

3) Write a number that is greater than 63.

_____ is greater than 63.

1.NBT.3

Do you understand?   ?   ✓

Affirmation: Answers are important, but questioning and solving are just as important.

Name: _____ Date: _____

Unit 9 Practice 10: I can write a question from a story problem.

4) Isabella is reading a book that has 125 pages. So far, Isabella has read 100 pages in her book.

Write a possible question for the story problem. _____

_____

_____

_____

_____

2.OA.1

5) Bryan runs every day. On Monday, Bryan ran for 64 minutes. On Tuesday, Bryan ran for 57 minutes.

Write a possible question for the story problem. _____

_____

_____

_____

_____

2.OA.1

Do you understand? ?

Affirmation: New solutions happen when we try new ways of doing things.

Name: _____   Date: _____

Unit 9 Practice 11: I can connect tape diagrams to story problems and equations.

1) Write the number that makes the equation true.

   35 + _____ = 76   68 - 33 = _____   54 + 27 = _____

   16 + _____ = 30   91 - 24 = _____   49 + 49 = _____

   23 + _____ = 62   46 - 37 = _____   52 + 58 = _____

   2.NBT.5

2) Write an equation that matches the tape diagram.

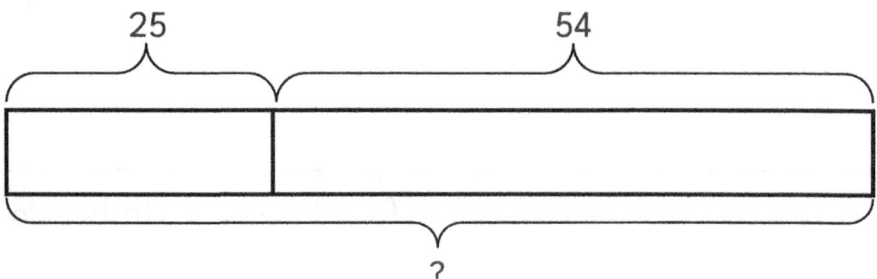

   Equation: _____

   2.OA.1

3) Write an equation that matches the tape diagram.

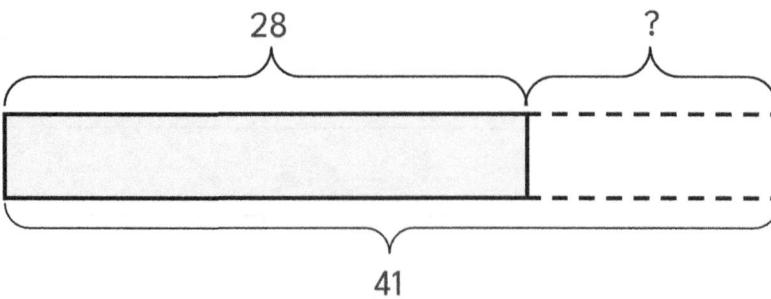

   Equation: _____

   2.OA.1

Do you understand?   ?   ✓

Affirmation: New solutions happen when we try new ways of doing things.

Name: _____   Date: _____

Unit 9 Practice 11: I can connect tape diagrams to story problems and equations.

4) Hudson has a large collection of mini cars. There are 24 red mini cars, 13 green mini cars and 25 blue mini cars in Hudson's collection. How many mini cars are in Hudson's collection? Draw a tape diagram to represent the story problem.

There are _____ mini cars in Hudson's collection.

Equation: _____

2.OA.1

5) Lindie made a shape design using 43 shapes. Twenty-seven of the shapes in the design are triangles and the rest are squares. How many squares are in Lindie's shape design? Draw a tape diagram to represent the story problem.

There are _____ squares in Lindie's shape design..

Equation: _____

2.OA.1

Do you understand?   ?

Affirmation: By asking questions, I learn more and help others learn as well.

Name: _____   Date: _____

Unit 9 Practice 12: I can solve and write one-step story problems.

1) Grayson and Willow collect action figures. Grayson has a collection with 32 action figures. Willow's collection has 19 action figures. How many more action figures are in Grayson's collection that in Willow's collection?
   Draw a tape diagram to represent the story problem.

   There are _____ more action figures in Grayson's collection.

   Equation: _____

   2.OA.1

2) Colton took some circles and some rectangles to a table in the art center. When spread out on the table, Colton noticed that there were more circles than rectangles. If Colton counted a total of 47 combined circles and rectangles, how many of the shapes on the table could be circles and how many of the shapes on the table could be rectangles?

   There could be _____ circles and _____ rectangles on the table.

   Equation: _____

   2.OA.1

Do you understand?   ?   ✓

Affirmation: By asking questions, I learn more and help others learn as well.

Name: _____ Date: _____

Unit 9 Practice 12: I can solve and write one-step story problems.

3) Write and solve a story problem that the tape diagram could represent.

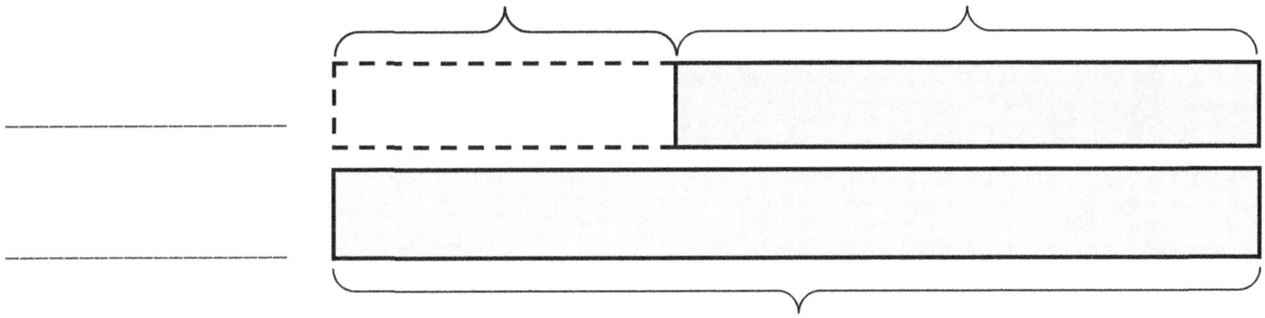

Story Problem: _____

_____

_____

_____

_____

Solve:

2.OA.1

Do you understand?   ?   ✓

Affirmation: I love learning! I am a learner.

Name: _____ Date: _____

Unit 9 Practice 13: I can solve and write two-step story problems.

1) Nolan has a sticker collection with 26 stickers. Nolan gets 52 stickers from the store to add to his collection. Nolan gives 18 of his stickers to his sister. How many stickers are in Nolan's collection now?

Nolan now has _____ stickers in his collection.

Equation: _____

2.OA.1

2) Layla bought 15 green apples and 16 red apples to make applesauce. Layla used 8 of the green apples and 2 of the red apples for a pie. How many apples combined did Layla have to make into applesauce?

Layla has _____ apples to make into applesauce.

Equation: _____

2.OA.1

Do you understand?   ?   ✓

Unit 9 Practice 13: I can solve and write two-step story problems.

3) Write and solve a story problem that the tape diagram could represent.

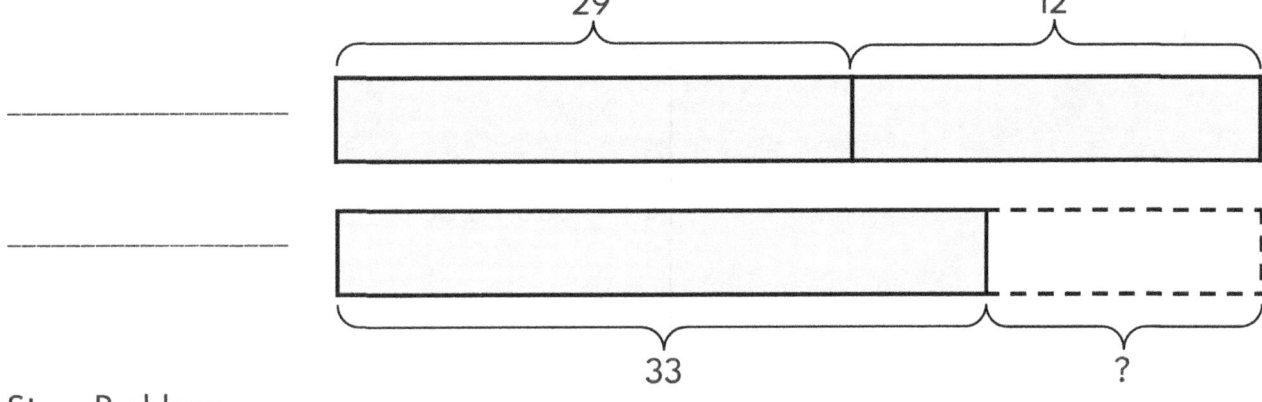

Story Problem: _____

_____

_____

_____

Solve:

2.OA.1

Do you understand?   ?   ✓

Affirmation: I get better the more I practice.

Name: _____ Date: _____

I can practice grade level fluencies.

Set 1: Find the sum with three addends

1) 2 + 2 + 2 = _____

2) 5 + 5 + 5 = _____

3) 4 + 6 + 7 = _____

4) 8 + 9 + 2 = _____

5) 3 + 1 + 7 = _____

6) 5 + 25 + 5 = _____

7) 7 + 7 + 13 = _____

2.NBT.6

Set 2: Find the sum or the difference

1) 216 + 132 = _____

2) 503 + 206 = _____

3) 341 + 341 = _____

4) 186 + 710 = _____

5) 930 - 820 = _____

6) 784 - 253 = _____

7) 689 - 73 = _____

2.NBT.5

Do you understand?   ?   ✓

Unit Reflection

Putting It All Together

Use this space to reflect on your understanding of the unit skills and concepts.

Skill/Concept	I can...	I need to work on...

Made in the USA
Middletown, DE
21 February 2026